ENTERPRISING MATHEMATICS

BY

THE SPODE GROUP

Oxford New York Tokyo
OXFORD UNIVERSITY PRESS
1986

Oxford University Press, Walton Street, Oxford OX2 6DP
Oxford New York Toronto
Delhi Bombay Calcutta Madras Karachi
Kuala Lumpur Singapore Hong Kong Tokyo
Nairobi Dar es Salaam Cape Town
Melbourne Auckland
and associated companies in
Beirut Berlin Ibadan Nicosia

Published in the United States
by Oxford University Press, New York

British Library Cataloguing in Publication Data
Spode Group
Enterprising mathematics.
1. Mathematics–Examinations, questions etc.
I. Title
510'.76 QA43
ISBN 0–19–853652–6

Library of Congress Cataloging in Publication Data
Main entry under title:
Enterprising mathematics.
1. Mathematics–Problems, exercises, etc.
I. Spode Group.
QA43.E66 1985 510'.76 85–28506
ISBN 0–19–853652–6 (pbk.)

Set by Cambrian Typesetters, Frimley, Surrey
Printed in Great Britain by
Butler and Tanner Ltd, Frome, Somerset

Contents

6. House design

7. Buying

8. Odds and ends

PART 2. SOLUTIONS TO PROBLEMS

Contributors

John Berry	Faculty of Mathematics, The Open University (Editor)
Roger Biddlecombe	Ounsdale High School, Staffordshire
Roger Blackford	Computing Advisory Teacher, Staffordshire
Morag Borrie	Pilgrim School, Bedford
David Burghes	School of Education, University of Exeter (Editor)
Bob Davison	School of Mathematics, Leicester Polytechnic
Bob Francis	Chester College of Higher Education
Ron Haydock	Mathematics Department, Matlock College
Ian Huntley	Department of Math Sciences, Sheffield City Polytechnic (Editor)
Robin Ingledew	Stainsby School, Middlesborough
Peter Moody	Mathematics Advisor, Dyfed
Graham Nellist	Boynton School, Middlesborough
Susan Pirie	Mathematics Education Centre, Warwick University
Paula Sellwood	Mathematics Advisory Teacher, Hertfordshire
John Walton	Monks Walk School, Welwyn Garden City, Hertfordshire
Barbara Young	Tarporley County High School, Cheshire

Illustrations

Nigel Weaver

The Spode Group

Director	Professor David Burghes
Associate Directors	Dr John Berry and Dr Ian Huntley
Secretary	Ms Sally Williams, School of Education, University of Exeter, St Luke's, Exeter EX1 2LU, UK

Introduction

As you probably won't want to use all the problems, we have left the choice to you and you are allowed to photocopy any parts of the text in Part 1 that you would like to use with your students.

We have arranged the contents in terms of topics – not mathematical topics but real-life topics – e.g. sports, money, travel, etc. We very much encourage you to teach mathematics in this way, so that your students will not feel that it is just another mathematics course.

At the beginning of each chapter, we have given you a short introduction, with ideas for class discussions on the topic. Again, we would encourage you to use discussions throughout. The idea throughout this text is to integrate mathematics into everyday life, and this will inevitably involve discussion, debate, and even argument! This will probably be very unlike your usual style of teaching mathematics, but don't worry. Mathematics doesn't have to be a tedious, arid subject – it can be controversial, entertaining, and enjoyable. We hope that this text will be of help to show this.

The second part of the book gives answers to the problems set in Part 1, although for many problems we cannot give precise answers.

We wish you success with your teaching and we very much hope that you will enjoy using this book, and that both you and your pupils will see the relevance of mathematics.

David Burghes
(Director – The Spode Group)

Part I. Text

1 Sport

It is not entirely obvious that mathematics and sport go together. Mathematics is not often thought of as a pastime or game, but it is surprising how often we use simple mathematics in sport. For example, in cricket the batting average is calculated by dividing the number of runs by the number of completed innings, i.e. by the use of *division*. In this chapter we investigate different sports and games which involve simple mathematical ideas in their solution.

Each of the sections contains sufficient details of the rules of the relevant game so that the problem can be solved. However, it might be a good idea to introduce the rules and objects of the games as a class activity to illustrate the features of the problems that need to be understood to make them tractable. For example, by playing the matchstick game of § 1.6 a winning strategy soon becomes apparent.

1.1. Super league football

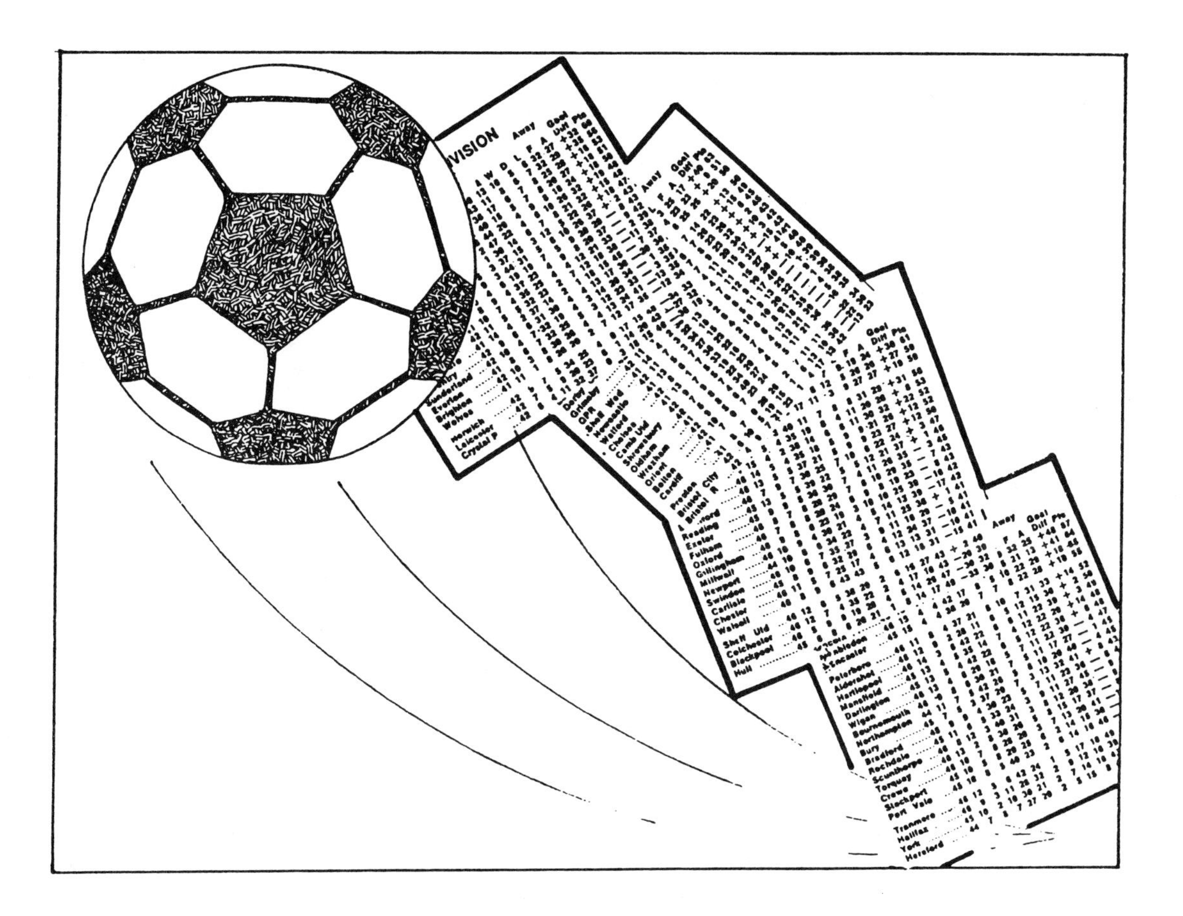

The first division of English football contains 22 teams. This means that 42 league matches are played each season, each team playing 21 others twice in the season. To this must be added the FA Cup and Milk Cup (Football League Cup) matches. The top six teams are also involved in European football competitions. This produces a very crowded fixture list. To reduce the load it was decided to split the first division into two unequal-sized divisions, with those involved in European football in the smaller group. Thus a super league was formed of six teams. The results of a season's matches, together with the attendance at each game, are given in Table 1.1.

Table 1.1

Home team	Score	Visitors	Score	Attendance	Home team	Score	Visitors	Score	Attendance
Aston Villa	3	Liverpool	1	42 371	Liverpool	1	Southampton	0	37 357
Ipswich	3	Aston Villa	2	24 671	Man City	1	Liverpool	3	47 281
Aston Villa	2	Tottenham	0	40 324	Liverpool	1	Tottenham	0	38 243
Southampton	0	Aston Villa	1	21 012	Man City	0	Ipswich	1	39 420
Aston Villa	2	Man City	2	39 587	Southampton	0	Man City	1	28 031
Liverpool	2	Aston Villa	2	41 253	Man City	2	Tottenham	3	35 624
Aston Villa	1	Ipswich	0	37 469	Ipswich	3	Man City	2	19 423
Tottenham	1	Aston Villa	3	46 290	Man City	1	Southampton	1	27 801
Aston Villa	1	Southampton	2	36 876	Tottenham	4	Man City	2	39 241
Man City	3	Aston Villa	0	34 421	Ipswich	2	Southampton	1	21 661
Liverpool	4	Ipswich	2	39 291	Tottenham	3	Ipswich	2	43 870
Southampton	1	Liverpool	2	26 775	Southampton	5	Ipswich	2	29 833
Liverpool	2	Man City	0	44 472	Ipswich	1	Tottenham	1	25 422
Tottenham	2	Liverpool	0	47 203	Southamptom	0	Tottenham	0	30 674
Ipswich	1	Liverpool	2	25 729	Tottenham	2	Southampton	0	39 783

● Problems

1. Complete the tables which follow for each of the teams. Calculate the number of points obtained by each team. A win earns three points and a draw earns one point. Aston Villa's chart has been completed as an example.

Aston Villa			Ipswich			Liverpool		
Win	Draw	Lose	Win	Draw	Lose	Win	Draw	Lose
3–1								
		2–3						
2–0								
1–0								
	2–2							
	2–2							
1–0								
3–1								
		1–2						
		0–3						
5 Won	2 Drawn	3 Lost	Won	Drawn	Lost	Won	Drawn	Lost
Points		17	Points			Points		
Goals for		17	Goals for			Goals for		
Goals against		14	Goals against			Goals against		

Manchester City			Southampton			Tottenham		
Win	Draw	Lose	Win	Draw	Lose	Win	Draw	Lose

Won	Drawn	Lost	Won	Drawn	Lost	Won	Drawn	Lost
Points Goals for Goals against			Points Goals for Goals against			Points Goals for Goals against		

2. Use the information you have gained in Problem 1 to make a league table. If two teams have the same points, then the team with the better goal difference is placed higher.

Name of team	Games				Goals		
	Played	Won	Drawn	Lost	For	Against	Points
Aston Villa	10	5	2	3	17	14	14

3. Complete the following tables to calculate the total attendance at each of the following grounds.

Home game	Aston Villa	Ipswich	Liverpool
1	42 371		
2	40 324		
3	39 587		
4	37 469		
5	36 876		
Total attendance	196 627		

Home game	Manchester City	Southampton	Tottenham
1			
2			
3			
4			
5			
Total attendance			

● Related problems

A. If the average admission charge at Liverpool is £2, how much did Liverpool earn for their five home games?

B. If the average admission charge at Ipswich is £2.20, how much did Ipswich earn for their five home games?

C. If the average admission charge at Tottenham is £2.50, how much did Tottenham earn for their five home games?

D. If Liverpool had raised their admission charges to an average of £2.20 per game, how much more would they have earned?

E. An Aston Villa player earns £100 bonus for every point scored. One player plays in all 10 games. How much bonus money will he have earned?

F. A Manchester City player earns £150 bonus for every point scored. One player plays in all 10 games. How much bonus money will he have earned?

G. A Liverpool player earns £120 bonus for every point scored. One player plays in every game. How much bonus money will he have received?

H. Southampton pays each of its players £100 bonus for each point scored. One player plays in eight matches. He misses both the matches against Tottenham. How much bonus money will he have earned?

I. Tottenham gives a bonus of £150 for a win and £50 for a draw. How much bonus will a player get if he plays in every game?

J. Ipswich are considering two systems of paying bonuses.

System 1: £180 for a win, and £60 for a draw;
System 2: £100 for each point.

How much would a player have earned in the 10 games if he had been paid under System 1?
How much would a player have earned in the 10 games if he had been paid under System 2?
Which system do you think he would prefer?

K. Devise a fixture list for the six teams playing on 10 Saturdays.

1.2. Snooker

The rules of snooker are probably familiar to you. If not, we will try briefly to explain them. The basic idea is to 'pot' all the balls (Fig. 1.1), which have values:

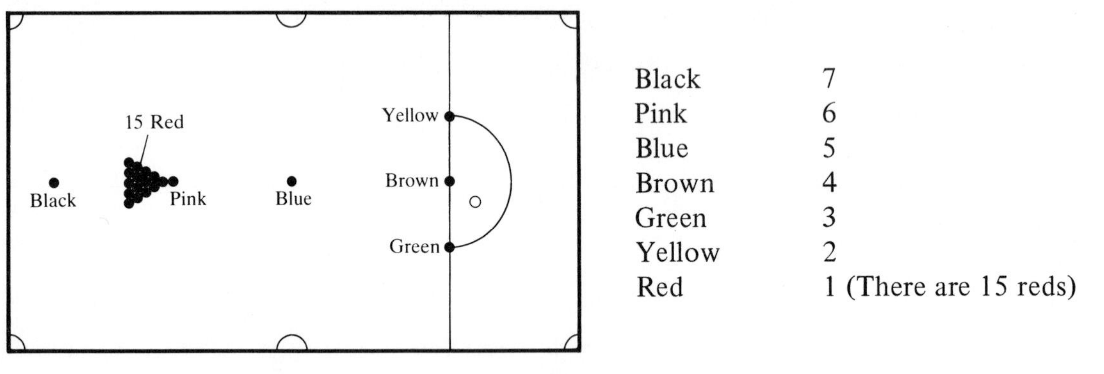

Fig. 1.1

Black	7
Pink	6
Blue	5
Brown	4
Green	3
Yellow	2
Red	1 (There are 15 reds)

The first competitor aims the white (the cue ball) from the 'D' area and has to pot a red followed by a colour; the colour is returned to its spot on the table and a red must next be potted, followed by another colour. This continues until all the reds have been potted. The colours must now be potted in order (yellow, green, brown, blue, pink, and black) and this time the colours are not returned.

Of course, it is very unlikely that the first player will go right through this sequence without failing. When he does fail in a shot, his score is recorded, and the second player takes over, who must start by potting a red. The play continues in this way, the score being added for each 'break', and the player with the highest total score when all the reds and colours have been potted is the winner.

● Problems

1. What is the maximum break that one player can make?
2. The score so far in a game is

Player A 7 Player B 48

Two reds and all the colours are left on the table. It is Player A's turn. Can he still win the game?

3. The score so far in a game is

Player A 28 Player B 52

No reds and all the colours are left on the table. It is the turn of Player A. Can he still win?

4. A player has arrived at the situation in Fig. 1.2, where there is only one red left to spot.

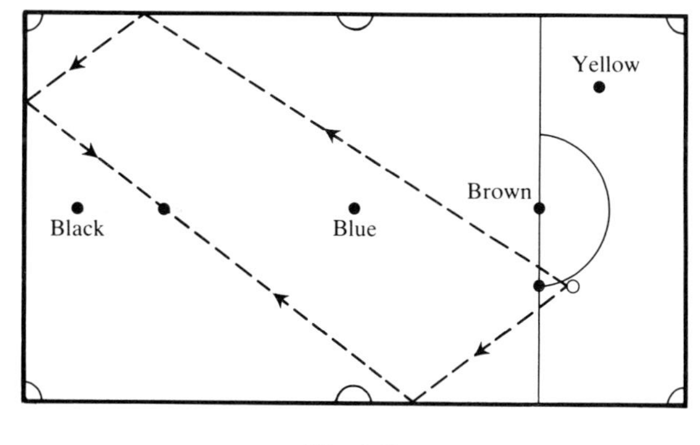

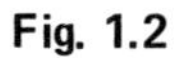
Fig. 1.2

Since the green is directly in the way, a snooker, it is not possible to hit the red without playing off the cushion on the side of the table. Two possible paths are indicated. Sketch some other possible paths. Find out how many possible paths there are if the cue ball is allowed to hit the cushion only once.

How many paths are there if the cue ball hits the cushion twice?

1.3. Darts

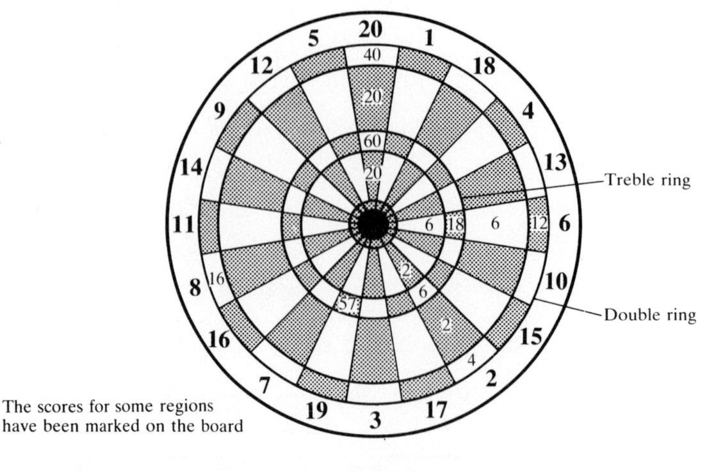

Fig. 1.3

Figure 1.3 shows the board used in most games nowadays, with doubles and trebles, bull and outer. If the dart lands in the outer ring, you score twice the number shown. If it lands in the next narrow ring, you score three times that number. In the centre the inner is the bull, worth 50, and the outer is worth 25. You throw three darts in each turn, from behind a line nine feet from the board called the ockey.

Suppose you're playing to score 501, finishing with a double or bull.

● Problems

1. You need 57 and it is your turn. How could you finish? There are a lot of answers. Start by finding five ways of finishing with three scoring darts, the last in the bull.

2. Again you need 57. With your first dart you throw 8. How many ways are there now of finishing?

3. Find the least number of throws needed to score 501 (ending with double or bull).

4. In Oxford, many boards have no treble wires. On such a board what is the least number of throws needed for 501?

5. Experts aim most of the time for treble 20, more often scoring 20. Occasionally they miss the 20 sector altogether, landing in 1 or 5. Suppose you are not very expert, so that if you aim for a sector you are just as likely to land in the next sector. In other words you are good enough not to miss by more than one sector you haven't yet achieved the accuracy of an Eric Bristow or a John Lowe. Does it make good sense for you to try for the 20? If not, where would be the best point on the board for you to aim, to build up your score? Do some calculations to explain your answers.

● Related problems

Now suppose you're organising a darts tournament, for individual players. It may be played over a whole season, so that there is time for everybody to play everybody else (an 'all-play-all' tournament). Let's take a look at a case with just a few players, first. Suppose four people compete, say A, B, C, and D. Then the matches are

$$A \times B, A \times C, A \times D, B \times C, B \times D, C \times D,$$

six games in all. Now try these.

Find how many matches there would be if there were five players, six players, . . . , 12 players. Can you find a quick way of working out the number of matches, without writing them all down? Suppose there were 20 competitors: how many matches?

1.4. Arranging a knockout tournament

In § 1.3 we looked at all-play-all (sometimes called 'American') tournaments. Unless the number of competitors is quite small, these tournaments take a long time. Suppose you wanted to stage a tournament in a single evening, to find a single winner. One possibility would be a knockout competition in which the losers in any round drop out and only the winners go forward. If there happens to be an odd number of players in any round then one must be given a bye. That is, in that round the player has no match but goes straight forward into the next round. Table 1.2 is a results table for a six-person knockout tournament. The competition has three rounds and there are five games.

Table 1.2

First round	Second round	Third round	(final)
Archer			
vs.	Bull		
Bull			
	vs.	Hood	
Flyte			
vs.	Hood		
Hood		vs.	Hood
Scorer			
vs.	Tell (bye)	Tell	
Tell			

● Problems

1. Instead of a bye in the second round in the six-person tournament, there could have been two byes in the first round. How many rounds then? And how many matches?

2. Draw up a table for an imaginary eight-person competition. How many rounds and how many matches?

3. Do the same for a nine-person competition. Does it matter where you place the byes? That is, does it change the number of rounds or matches?

4. Have you found a rule giving the number of rounds and the number of matches? Give the answers for a tournament with

(a) 14; (b) 21; (c) 33 competitors.

● Related problems

Knockout competitions are sometimes criticized because a single freak result can rob a good player of the chance to go forward. If there is more time than the bare minimum needed for a knockout tournament, but not enough for an American tournament, a 'Swiss' tournament may be played. In a Swiss tournament, at the end of each round the organizers pair off competitors with (as nearly as possible) the same total score to date. This means that players of about equal ability should tend to meet in the later rounds. About six rounds usually produce a fair result if the number of competitors is less than 20.

For a 12-person competition, find the number of games played in

(a) an American tournament;
(b) a Swiss tournament;
(c) a knockout tournament.

Think of one respect in which a Swiss tournament is less effective than a knockout tournament, and one way in which it is better. How would you remedy the weakness of the Swiss tournament?

1.5. Cribbage

In the game of cribbage each player is dealt four cards. Points are awarded for any of the following.

1. Totals of 15 (2 points);
2. Runs of 3, 4, or 5 cards (1 point per card);
3. Pairs of cards of equal value (2 points);
4. 4 or 5 cards of the same suit, known as a flush (1 point per card).

A fifth card is turned up and can be used by each player to score additional points.

In this game an ace counts 1 and all picture cards count 10.

Suppose a player is dealt the 4 hearts (4H), 5 clubs (5C), 6 hearts (6H), and queen diamonds (QD). The turn-up card is the jack of clubs (JC).

Totals of 15 can be made using

(i) 4H + 5C + 6H
(ii) 5C + QD
(iii) 5C + JC

and 2 points is scored for each 15.

A run of the 4H, 5C, 6H scores an additional 3 points, giving a total score of 9 points (6 for 15s and 3 for the run).

A hand consisting of the 7 hearts (7H), 7 clubs (7C), 8 hearts (8H), and 9 hearts (9H) is now dealt. The turn-up card is the 6 hearts (6H). On this occasion it is possible to score points for totals of 15, pairs of equal value, and runs. (Note that a flush is not possible here because to make a flush all the cards in the originally dealt hand (7H, 7C, 8H, 9H) must be of the same suit.)

The combinations and points scored are illustrated in Table 1.3.

Table 1.3

	Combinations	Points scored
15s	7H + 8H	2
	7C + 8H	2
	6H + 9H	2
Pair	7H, 7C	2
Runs	6H, 7H, 8H, 9H	4
	6H, 7C, 8H, 9H	4
Total		16

Now consider a hand consisting of the 8 hearts (8H), 3 clubs (3C), 3 diamonds (3D), and 5 clubs (5C) with a turn-up card 6 hearts (6H) (Table 1.4).

Table 1.4

	Combinations	Points scored
15s	3H + 3C + 3D + 6H	2
Runs	3H, 3C	2
	3H, 3D	2
	3C, 3D	2
Total		8

If you still find difficulty with the scoring system turn to the extra material at the end of this section for further help before attempting the problems.

● Problems

1. Calculate the number of points scored with each hand:

(i)	Hand	2 spades, 4 clubs, 10 hearts, 8 clubs
	Turn-up	3 diamonds
(ii)	Hand	King diamonds, 8 diamonds, 10 clubs, 5 hearts
	Turn-up	5 diamonds
(iii)	Hand	2 clubs, 10 clubs, 9 clubs, jack clubs
	Turn-up	3 hearts
(iv)	Hand	3 spades, ace spades, 9 clubs, 5 hearts
	Turn-up	4 clubs
(v)	Hand	ace hearts, 7 clubs, 7 diamonds, 6 spaces
	Turn-up	2 spades
(vi)	Hand	queen clubs, 6 clubs, ace hearts, 9 spades
	Turn-up	8 clubs
(vii)	Hand	ace diamonds, 8 clubs, queen spades, 8 spades
	Turn-up	4 clubs
(viii)	Hand	9 diamonds, 9 spades, queen hearts, queen clubs
	Turn-up	7 spades

(ix) Hand 6 spades, 9 spades, 2 spades, 7 spades
Turn-up 8 spades

(x) Hand jack hearts, 7 hearts, 6 diamonds, queen clubs
Turn-up 4 clubs

2. The rules of the game are now changed so that each player is dealt five cards of which one must be thrown away.

Suppose the cards dealt were the 3 spades (3S), 5 spades (5S), 7 hearts (7H), 8 clubs (8C), and the ace diamonds. Which is the best card to throw away?

Consider discarding each card in turn.

	Discarded card									
	3 spades		5 spades		7 hearts		8 diamonds		Ace diamonds	
	Combinations	Points	Combinations	Points	Combinations	Points	Combinations	Points	Combinations	Points
15s	7H+8D	2	7H+8H	2	–	0	3S+5S+7H	2	3S+5S+7H 7H+8C	2 2
Total		2		2		0		2		4

Discarding the ace diamonds gives the best points total. The points total can be improved by all turn-ups with the exception of an ace. The best turn-up card is a four.

A player is now dealt the king diamonds (KD), 9 diamonds (9D), 6 diamonds (6D), 10 diamonds (10D), and the 9 spades (9S). Which is the best card to throw away on this occasion?

Consider discarding each card in turn.

	Discarded card									
	king diamonds		9 diamonds		6 diamonds		10 diamonds		9 spades	
	Combinations	Points	Combinations	Points	Combinations	Points	Combinations	Points	Combinations	Points
15s	9D+6D 9S+6D	2 2	9S+6D	2	–	0	9D+6D 9S+6D	2 2	9D+6D	2
Pair	9D, 9S	2	–	0	–	0	9S, 9D	2	–	0
Flush	–	0	–	0	–	0	–	0	KD, 9D, 6D, 10D	4
Total		6		2		0		6		6

In this situation it is necessary to consider the possible turn-up cards to give additional points to determine which is the best combination.

A player is now dealt five cards and one card must be thrown away. The five cards he was dealt were the 5 hearts, queen spades, jack spades, 6 diamonds, and the 10 clubs. Which is the best card to throw away?

Consider discarding each card in turn.

	Discarded card									
	5 hearts		queen spades		jack spades		6 diamonds		10 clubs	
	Combinations	Points	Combinations	Points	Combinations	Points	Combinations	Points	Combinations	Points
15s	–	0	5H+JS 5H+10C	2 2	5H+QS 5H+10C	2 2	5H+QS 5H+JS 5H+10C	2 2 2	5H+QS 5H+JS	2 2
Run	–	0	–	0	–	0	10C, JS, QS	3	–	0
Total		0		4		4		9		4

Discarding the 6 diamonds gives the best points total. If the card turned up is a spade, an extra point is scored for having the jack of that suit in the hand.

For further help turn to the extra work before attempting Problem 3.

3. Find the best card to throw away in each of the following deals.

(i) king spades, king hearts, jack hearts, 7 clubs, 7 diamonds;
(ii) 6 spades, 6 diamonds, 4 spades, 9 diamonds, 9 clubs;
(iii) 8 clubs, queen clubs, ace clubs, king clubs, jack diamonds
(iv) 7 hearts, queen diamonds, 4 diamonds, 8 hearts, 7 spades;
(v) 10 hearts, 10 diamonds, ace diamonds, 9 diamonds, 4 hearts;
(vi) queen clubs, king spades, 4 spades, 9 hearts, 5 hearts;
(vii) jack spades, 8 diamonds, 7 hearts, ace spades, 9 spades;
(viii) 6 spades, 4 hearts, 7 clubs, 9 clubs, 6 diamonds;
(ix) 3 clubs, 2 diamonds, queen hearts, 6 hearts, 10 diamonds;
(x) 7 diamonds, 2 clubs, ace diamonds, 8 hearts, 7 spades.

The game can now be played involving two or more players. Five cards are dealt to each player who throws away one card of his or her choice. A card is turned up and each player keeps a record of his or her scores. The first player to reach a total of 50 points is the winner.

● Related problems

A. A player is dealt the hand of 9 hearts, queen diamonds, 6 hearts, and 3 clubs. The card turned up is the 6 clubs. How many points are scored?

	Combinations	Points
15s	9H + 6H	
	9H + 6C	
	6H + 3C + 6C	
Pair	6H, 6C	
Total		

B. A player is dealt the hand ace clubs, 10 hearts, ace hearts, and 4 spades. The card turned up is the 3 spades. How many points are scored?

	Combinations	Points
15s		2
		2
		2
Pair		2
Total		

C. A player is dealt the hand 2 clubs, 3 clubs, 4 diamonds, and king hearts. The card turned up is the 6 spades. How many points are scored?

	Combinations	Points
15s		
Run		3
Total		

D. A player is dealt the hand 4 hearts, 5 clubs, 6 hearts, and the jack spades. The card turned up is the 5 diamonds. How many points are scored?

	Combinations	Points
15s		
Run		
Pair		
Total		

E. A player is dealt the hand 3 clubs, 3 hearts, 3 diamonds, and 2 hearts. The card turned up is the king spades. How many points are scored?

	Combinations	Points
15s		
Pairs		
Total		

F. A player is dealt the hand 4 clubs, 2 diamonds, 6 diamonds, 7 clubs, and 3 spades. Which is the best card to throw away?

Consider throwing away each card in turn.

	Discarded card									
	4 clubs		2 diamonds		6 diamonds		7 clubs		3 spades	
	Combi-nations	Points	Combi-nations	Points	Combi-nations	Points	Combi-nations	Points	Combi-nations	Points
15s	2D+6D+7C						4C+2D+6D+ 3S		2D+6D+ 7C	
Run					2D, 3S, 4C		2D, 3S, 4C			
Total										

The best card to throw away is the 7 clubs.

G. A player is dealt the hand 6 spades, 7 clubs, 8 hearts, 2 spades, and the 5 diamonds.

Consider throwing away each card in turn.

	Discarded card									
	6 spades		7 clubs		8 hearts		2 spades		5 diamonds	
	Combi-nations	Points	Combi-nations	Points	Combi-nations	Points	Combi-nations	Points	Combi-nations	Points
15s		2 2		2		2		2		2 2
Run						3		4		3
Total										

Which is the best card to throw away?

1.6. Matchstick men

In this game for two players, 11 matches are laid out on the table. The players take it in turns to remove 1, 2, or 3 matches from those on the table, and the winner is the person who takes the last match.

Play a few times, and see if you can devise a winning strategy. Do you need to start in order to win?

Now read on.

Since you are only allowed to remove 1, 2, or 3 matches, the winning strategy is always to leave a multiple of 4 matches on the table. Try this and see how it works.

If you are playing with someone who knows the strategy, you need to start in order to win. Otherwise you need, as soon as possible, to get to a situation where a multiple of 4 is left.

● Related problems

A. Now try with 21 matches, instead of 11. Can you still win?

B. What happens if you can take 2, 3, or 4 matches away from 21?

C. What if you can take 4, 5, or 6 matches from a total of 35?

D. What if you can take 4, 5, 6, or 7 matches from 35?

E. Can you work out a general strategy whatever the size of the heap and whatever the choice of number to take away?

1.7. Bets with a die

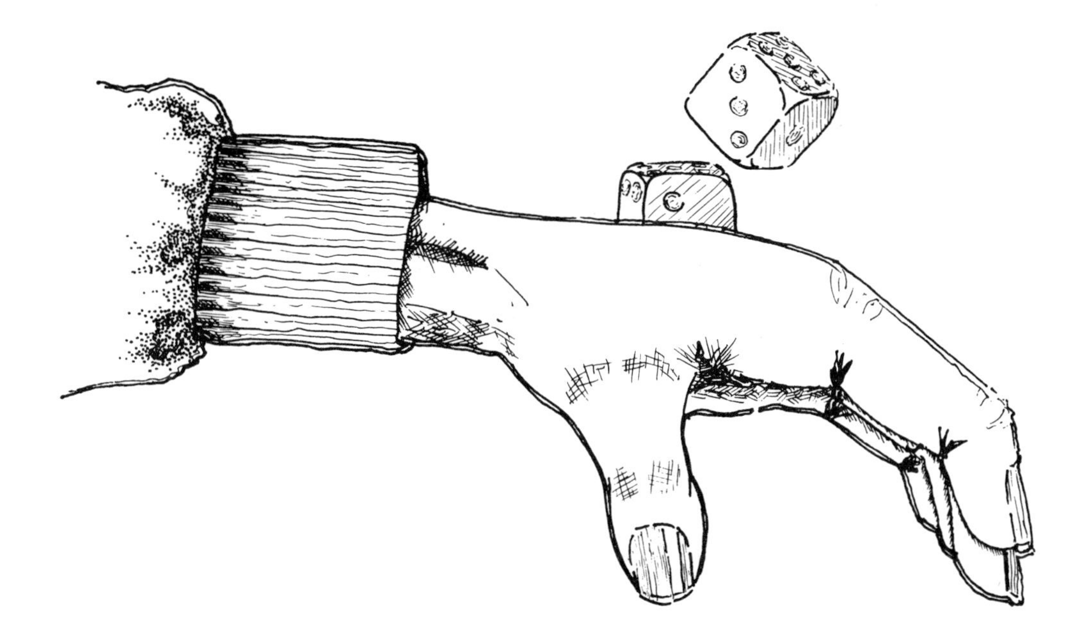

A friend says 'I'll give you £1 if you roll this die once and throw a one – anything else and you give me £1.' Is this bet worth taking?

No! There is only one throw which wins, but there are five throws which lose (two, three, four, five, and six). Since there is an equal chance of any of 1, 2, 3, 4, 5, or 6 being thrown, there is a 1 in 6 chance of winning and 5 in 6 chance of losing – this means that, on average, for every six throws, you will win on 1 and lose on 5. Clearly not a good bet!

The bet would be worth taking only if the number of winning throws is greater than the number of losing throws.

● Problems

1. The friend now says 'If you throw three, four, five, or six with one go, I'll give you £1. If not you give me £1'. Is this bet worth taking?

2. The friend now says 'If you throw one or two with one go, I'll give you £2. If not you give me £2'. Is this bet worth taking?

The problem gets more difficult if we change the amounts of money for winning and losing. For example, if your friend now says, 'If you throw one or two I'll give you £3; if not you give me £1', is this bet worth taking?

We can find out by considering what will happen on an average six throws of the die. Twice you will win and the sum of money won is

$$2 \times £3 = £6.$$

Four times you will lose, and the sum of money lost is

$$4 \times £1 = £4.$$

Thus, on an average six throws, you will win

$$£6 - £4 = £2$$

and this bet is worth accepting.

● Related problems

For each of the bets below, find out if the bet is worth taking.

A. 'You win £5 if you throw a 3 or 4, and I win £2 for anything else.'

B. 'You win £12 if you throw a 6, and I win £5 for anything else.'

C. 'You win £4 if you throw a 1, 2, 3, or 4 and I win £8 for anything else.'

2 Travel

To many families, the summer holiday is one of the high points in the year. The planning often begins in early January, coinciding with the many television advertisements tempting the viewers to sunny places. There are many aspects to planning a holiday. For example, an important decision is whether to travel abroad or to take a holiday in Britain. But there are ways of spending a holiday period other than 'going away'. For example, British Rail offer special railrover tickets for unlimited travel within a certain area. This gives an opportunity to travel locally to visit the 'local places of interest'.

When planning a holiday, the cost is an important feature. There are often 'hidden costs' such as airport landing fees. This chapter investigates various problems associated with travelling. Two of the sections investigate the costs of taking a holiday overseas, while in the third section we take to the train.

Here are some discussion points which may stimulate a class discussion and highlight some of the aspects that are needed to proceed with the problems.

1. What features are involved with planning a holiday abroad?
2. Is it cheaper to take a holiday in Britain?
3. The idea of a toll to use a road may be unfamiliar.
4. Which European countries have road tolls?

For § § 2.3 and 2.4, it is important that a British Rail (BR) timetable can be read and interpreted.

2.1. A holiday abroad

Table 2.1 shows details of the costs of a package holiday to Malta, taken from a travel brochure. The following details are also relevant.

Children aged 11 or over are charged at the adult rate.

There will be a reduction of 10 per cent for one child if all the family use one room.

Airport charges of £9.20 per person are added to the bill.

Mr Smith and his wife and two children, aged 6 and 13, want to go on holiday to Malta. After reading the travel brochures they decide that they would like to stay in one of the three hotels listed in Table 2.1 below. If they want to start their holiday between 9th August and 22nd August and stay for seven nights, work out the cost of a holiday in the hotels Betiana and Galien. The holiday in the hotel Alphina has been done for you.

Hotel Alphina	
2 adults at £220	£440.00
1 child at adult cost of £220	£220.00
1 child at £154 less 10 per cent	
10 per cent of 154	
10/100 × £154 = £15.40	
£154.00 – £15.40	£138.60
Airport charges for 4 people	
£9.20 × 4	36.80
Total	£835.40

Now work out the cost for the other two hotels.

Table 2.1

Basic holiday prices per person in £'s
See panel on right for departure airports and any applicable supplements

Hotel	Alphina (half board)				Betiana (half board)				Galien (half board)			
No. of nights	7 nights		14 nights		7 nights		14 nights		7 nights		14 nights	
	Adult	Child	Adult	Child	Adult	Child	Adult	Child	Adult	Child	Adult	Child
1 Apr–30 Apr	–	–	–	–	–	–	–	–	179	110	249	122
1 May–13 May	193	125	249	137	198	127	259	139	219	129	296	141
14 May–20 May	195	128	251	140	199	130	261	142	221	131	299	143
21 May–27 May	197	131	254	144	202	133	264	146	224	133	302	145
28 May–20 Jun	199	134	257	148	204	136	266	150	226	135	304	147
21Jun–27 Jun	202	137	268	151	206	139	278	153	228	137	306	149
28 Jun–8 Jul	210	140	278	154	214	142	287	156	230	139	308	151
9 Jul–22 Jul	213	150	284	164	218	152	294	166	232	142	310	154
23 Jul–8 Aug	224	159	298	174	232	164	309	176	243	152	321	167
9 Aug– 22 Aug	220	154	292	168	226	158	304	170	237	145	315	157
23 Aug–31 Aug	214	151	284	164	222	154	294	166	232	142	311	154
1 Sep–18 Sep	212	144	280	157	219	148	290	160	230	140	308	152
19 Sep–26 Sep	209	141	270	155	216	144	279	158	227	136	305	148
27 Sep–6 Oct	197	136	254	150	203	141	264	154	223	133	301	145
7 Oct–23 Oct	194	131	251	143	201	136	261	147	220	131	297	143

For conditions applicable to child prices see page 4
For holidays including Good Friday and/or Easter Monday add £10

● Problems

1. In the hotel which you found to be the cheapest, what would be the cost of a 7-day holiday for the same family, taken between 1st May and 13th May. What is the saving?

2. How much more expensive is it to take a 14-day holiday for the family choosing to travel in early May?

3. Using travel brochures, pick a holiday of your choice and work out the cost. *Hint*: take care to read conditions at front of brochures and the small print.

● Related problems

A. Mr and Mrs Jones and their 2 children, aged 9 and 13, who live in Birmingham are interested in a holiday in the Swiss lake district. They looked through a holiday brochure and Hotel Delta caught their eye (Fig. 2.1). They decided to estimate the cost of a holiday from 1st June for one week. They realized they had to include the cost of travel to the airport, supplements, and insurance as well as airport charges. The reduction for children is calculated differently in the brochure. How much will the holiday cost?

Extracts from 'Spode Summer Fun' brochure 1982

All prices in £ per person	DELTA Half board			
From	Luton		Gtwk	B'ham
No. of nights	7	14	14	14
Take-off time	Tues 1600	Tues 1800	Sun 0815	Sun 1525
Home-landing	Tues 2050	Tues 2050	Sun 1905	Mon 0220
2 May–12 May	142	203	215	215
13 May–20 May	150	214	226	226
21May–31 May	159	227	239	239
1 June–11 June	155	223	235	235
12 June–22 June	161	229	241	241
23 June–4 July	168	238	250	250
5 July–16 July	171	243	255	255
17 July–23 July	180	252	264	264
24 July–15 August	185	259	271	271
16 Aug–30 August	174	241	253	253
31 Aug–14 September	167	228	240	240
15 Sept–21 September	160	–	–	–
First departure	4 May	4 May	2 May	2 May
Last departure	21 Sep	14 Sep	12 Sep	12 Sep

LAKE GARDA
Hotel Delta

Position: This newly completed hotel is situated about 200 yards from the lake shore and its attractive lido set back from the main road in wooded countryside provides beautiful peaceful surroundings for a restful holiday. Conveniently placed for local entertainment, shopping, and restaurants.

- attractive swimming pool
- bar, 2 restaurants
- lift
- pony treks into the hills

Children's reduction: 2nd cat.
Prices shown are per person for half board in a twin-bedded room, with private shower and WC.

Children's Reductions

This year we are offering our biggest children's reductions ever, in many cases your children can go free, and for the first time we are now offering reductions for children up to 16 years old.

- **2 to under 12 can often go free or in many cases get up to 50% reduction**
- **12 to under 16 often get 10% reduction**
- **1 child sharing with 1 adult often gets a 10% reduction**
- **Under 2's travel free**

At many hotels we have negotiated specially favourable children's reductions. These are indicated by the children free symbol on the hotel page. In a few cases where hoteliers do not give adequate reductions, this is shown in the relevant hotel descriptions.

All reductions are on the basic price shown in the price panels. The table sets out reductions for children 2 to under 12.

	1st Cat. Red.	2nd Cat. Red.	3rd Cat. Red.
Up to 30 April*	15%	15%	15%
1 May–20 May	Free	Free	50%
21 May–11June	Free	50%	25%
12 June–22 June	50%	25%	10%
23 June–4 July	50%	10%	10%
5 July–16 July	25%	10%	10%
17 July–14 Sept	10%	10%	10%
15 Sept–30 Sept	50%	25%	10%
1 Oct–15 Oct	Free	Free	10%
16 Oct–31 Oct	15%	10%	10%

FOR CHILDREN 12 TO UNDER 16 a reduction of 10% on the adult basic price will apply when sharing a room except for departures prior to the first of May and for departures between 21st July to 31st August inclusive.

UNDER 2's TRAVEL FREE

Airport Charges

Airport charges are becoming increasingly complicated with different charges being imposed for different times of the year and also the day of the week, departure times, etc. To simplify matters we have decided to pay them for you and add an average price of £9.45 to your final account. This includes estimated UK passenger charges, foreign airport taxes and UK airport security charges together with a small administration charge. Airport charges and taxes can be changed by Local Authorities or Governments at short notice. Should this happen, we reserve the right to pass on any increases within the terms of our price guarantee and Fair Trading Bond.

INSURANCE PREMIUMS

Up to 8 nights holiday	£5.95
9–16 nights holiday	£7.30
17–23 night holiday	£8.90

Fig. 2.1

Try this for yourself. Refer to the solution at the end of this section if you are having problems.

B. What is the cost of the holiday if they went from 13th to 20th May?

C. If they decide to have a 2-week holiday (at the same hotel) staying the last week of May and first week in June what will the total cost be? (*Note*: Package charges vary from week to week and for a 2-week holiday they can fly from Birmingham. What is the insurance charge?)

D. Collect your own brochures. Choose a holiday that appeals to *you* and work out the cost.

E. Can you think of an easier way of representing the information in the brochure?

2.2. Motoring to France

An increasing number of people travel to the Continent for their holidays, and many take their car for the freedom, convenience, and economy it provides.

The cost involved depends on several factors, such as

1. Cost of travel to channel port (this is affected by the type of car, distance involved, and cost of any overnight stops).
2. Cost of ferry crossing (varies according to ferry company, number in party, length of car, whether trailer/caravan is being towed, tariff for time of year/time of day, whether berths/cabins are required).
3. Cost of travelling to destination on the Continent (depends on the factors mentioned in (1) above, and also the cost of petrol abroad and any road toll charges – these will apply if motoring on the main motorways, or Autoroutes).
4. Insurance costs (to cover vehicle breakdown, baggage loss, medical expenses, cancellation costs).

Suppose two adults wish to travel from Bristol to Paris in a BL Allegro, via Southampton and Le Havre, sailing near mid-day on a Saturday in mid-July with P & O Ferries.

Reference to Tables 2.2–2.5 gives the following data.

Table 2.2. Mileage guide: UK (miles)

	Dover	Portsmouth	Southampton	Weymouth
Bristol	198	97	75	
London	77			
Manchester	283		227	238

Table 2.3. Mileage guide: France (miles)

	Calais	Cherbourg	Le Havre
Bordeaux	525	400	
Paris	172	221	115
La Rochelle			320

Table 2.4. Car data

	Length (m)	m.p.g*
BL Allegro	3.9	35
Vauxhall Cavalier	4.4	32
Ford Cortina	4.3	25
Ford Escort	3.97	36

* Assume that towing a caravan reduces the m.p.g. by 8 and that a trailer reduces it by 2. Take the price of petrol as £2 per gallon.

Table 2.5. Crossing times

	Dover–Calais	Dover–Boulogne
Hoverspeed	35 min	
P & O Ferries		1 hr 40 min
Sealink	1 hr 30min	
Townsend-Thoresen	1 hr 15 min	

Distance from Bristol to Southampton – 75 miles
Distance from Le Havre to Paris – 115 miles
Allegro: fuel consumpton 35 m.p.g
length 3.9 m

In the 1982 P & O brochure, the ferry tariff for 1100 hours sailing, July 10th, for a vehicle of length 3.9 m, driver and 1 adult being £68.

$$\text{Total number of gallons used} = \frac{(75 + 115)}{35} = 5.4$$

$$\text{Total cost of petrol} = 5.4 \times £2 = £10.80$$

The total travel cost is £68 + £10.80 = £78.80.

Portsmouth-Le Havre

Approximate sea crossing time 5½ hours day sailings.
2200, 2300 and 2330 arrive 0700 next day.

DEPARTURE DATES OF SAILING ALL LOCAL TIMES 1985

TIMES	T	W	T	F	S	S	M	T	W	T	F	S	S	M	T	W	T	F	S	S	M	T	W	T	F	S	S	M	T	W	T
JAN	1	2	3	4	5	6	7	8	9	10	11	12	13	14	15	16	17	18	19	20	21	22	23	24	25	26	27	28	29	30	31
1100	E	E	E	E	E	E	E	E	E	E	E	E	E	E	E	E	E	E	E	E	E	E	E	E	E	E	E	E	E	E	E
1500	E	E	E	E	E	E	E	E	E	E	E	E	E	E	E	E	E	E	E	E	E	E	E	E	E	E	E	E	E	E	E
2300	D	D	D	D	D	D	D	D	D	D	D	D	D	D	D	D	D	D	D	D	D	D	D	D	D	D	D	D	D	D	D
	F	S	S	M	T	W	T	F	S	S	M	T	W	T	F	S	S	M	T	W	T	F	S	S	M	T	W	T			
FEB	1	2	3	4	5	6	7	8	9	10	11	12	13	14	15	16	17	18	19	20	21	22	23	24	25	26	27	28			
0900																										E	E	E			
1100	E	E	E	E	E	E	E	E	E	E	E	E	E	E	E	E	E	E	E	E	E	E	E	E	E	E	E	E			
1500	E		E	E	E	E	E	E	E	E	E	E	E	E	E	E	E	E	E	E	E	E	E	E	E	E	E	E			
2200																									D	D	D	D			
2300	D	D	D	D	D	D	D	D	D	D	D	D	D	D	D	D	D	D	D	D	D	D	D	D	D	D	D	D			
	F	S	S	M	T	W	T	F	S	S	M	T	W	T	F	S	S	M	T	W	T	F	S	S	M	T	W	T	F	S	S
MAR	1	2	3	4	5	6	7	8	9	10	11	12	13	14	15	16	17	18	19	20	21	22	23	24	25	26	27	28	29	30	31
0900	E	E	E	E	E	E	E	E																							
1100	E	E	E	E	E	E	E	E	E	E	E	E	E	E	E	E	E	E	E	E	E	E	E		E	E	E	E	E	E	E
1500	E	E	E	E	E	E	E	E	E	E	E	E	E	E	E	E	E	E	E	E	E	E	E	E	E	E	E	E	E	E	E
2200	D	D	D	D	D	D	D	D																							
2300	D	D	D	D	D	D	D	D	D	D	D	D	D	D	D	D	D	D	D	D	D	D	D	D	D	D	D	D	D	D	D
	M	T	W	T	F	S	S	M	T	W	T	F	S	S	M	T	W	T	F	S	S	M	T	W	T	F	S	S	M	T	
APR	1	2	3	4	5	6	7	8	9	10	11	12	13	14	15	16	17	18	19	20	21	22	23	24	25	26	27	28	29	30	
1100	E	E	E	D	E	E	E	E	E	E	E	E	E	E	E	E	E	E	E	E	E	E	E	E	E	E	E	E	E	E	
1500	E	E	E	D	E	E	E	E	E	E	E	E	E	E	E	E	E	E	E	E	E	E	E	E	E	E	E	E	E	E	
2300	D	D	D	C	D	D	D	D	D	D	D	D	D	D	D	D	D	D	D	D	D	D	D	D	D	D	D	D	D	D	
	W	T	F	S	S	M	T	W	T	F	S	S	M	T	W	T	F	S	S	M	T	W	T	F	S	S	M	T	W	T	F
MAY	1	2	3	4	5	6	7	8	9	10	11	12	13	14	15	16	17	18	19	20	21	22	23	24	25	26	27	28	29	30	31
1100	E	E	E	E	E	E	E	E	E	E	E	E	E	E	E	E	E	E	E	E	E	E	E	C	D	D	E	E	E	E	E
1500	E	E	E	E	E	E	E	E	E	E	E	E	E	E	E	E	E	E	E	E	E	E	E	C	D	D	E	E	E	E	E
2200																							D	B	D	D	D	D	D	D	D
2300	D	D	D	D	D	D	D	D	D	D	D	D	D	D	D	D	D	D	D	D	D	D									
	S	S	M	T	W	T	F	S	S	M	T	W	T	F	S	S	M	T	W	T	F	S	S	M	T	W	T	F	S	S	
JUN	1	2	3	4	5	6	7	8	9	10	11	12	13	14	15	16	17	18	19	20	21	22	23	24	25	26	27	28	29	30	
0830																											E	D	D	D	
1100	E	E	E	E	E	E	E	E	E	E	E	E	D	C	D	E	E	E	E	E	E	E	E	E	E	E					
1500	E	E	E	E	E	E	E	E	E	E	E	E	D	C	D	E	E	E	E	E	E	E	E	E	E	E	E	D	D	D	
2200	D	D	D	D	D	D	C	D	D	D	D	D	C	B	D	D	D	D	D	D	C	D	D	D	D						
2330																										D	D	C	D	D	
	M	T	W	T	F	S	S	M	T	W	T	F	S	S	M	T	W	T	F	S	S	M	T	W	T	F	S	S	M	T	W
JUL	1	2	3	4	5	6	7	8	9	10	11	12	13	14	15	16	17	18	19	20	21	22	23	24	25	26	27	28	29	30	31
0830	D	D	D	D	C	C	D	D	D	D	D	C	C	D	D	D	D	C	B	B	C	C	C	C	C	B	B	C	C	C	C
1500	D	D	D	D	D	D	D	D	D	D	D	C	C	D	D	D	D	C	C	C	C	C	C	C	C	C	C	C	C	C	C
2330	C	C	C	C	C	C	C	C	C	C	C	B	C	C	C	C	C	B	B	C	C	C	C	C	B	B	C	C	C	C	C
	T	F	S	S	M	T	W	T	F	S	S	M	T	W	T	F	S	S	M	T	W	T	F	S	S	M	T	W	T	F	S
AUG	1	2	3	4	5	6	7	8	9	10	11	12	13	14	15	16	17	18	19	20	21	22	23	24	25	26	27	28	29	30	31
0830	C	B	B	C	C	C	C	C	C	C	C	C	C	C	C	C	C	D	D	D	D	D	D	D	D	D	D	D	D	D	D
1500	C	C	C	C	C	C	C	C	C	C	C	C	C	C	C	C	C	D	D	D	D	D	D	D	D	D	D	D	D	D	D
2330	B	B	C	C	C	C	C	C	B	C	C	C	C	C	C	B	C	C	C	C	C	C	C	C	C	C	C	C	C	C	C
	S	M	T	W	T	F	S	S	M	T	W	T	F	S	S	M	T	W	T	F	S	S	M	T	W	T	F	S	S	M	
SEP	1	2	3	4	5	6	7	8	9	10	11	12	13	14	15	16	17	18	19	20	21	22	23	24	25	26	27	28	29	30	
0830	E																														
1100		E	E	E	E	E	E	E	E	E	E	E	E	E	E	E	E	E	E	E	E	E	E	E	E	E	E	E	E	E	
1500	E	E	E	E	E	E	E	E	E	E	E	E	E	E	E	E	E	E	E	E	E	E	E	E	E	E	E	E			
1600																													E	E	
2300	D	D	D	D	D	D	D	D	D	D	D	D	D	D	D	D	D	D	D	D	D	D	D	D	D	D	D	D	D	D	
	T	W	T	F	S	S	M	T	W	T	F	S	S	M	T	W	T	F	S	S	M	T	W	T	F	S	S	M	T	W	T
OCT	1	2	3	4	5	6	7	8	9	10	11	12	13	14	15	16	17	18	19	20	21	22	23	24	25	26	27	28	29	30	31
1100	E	E	E	E	E	E	E	E	E	E	E	E	E	E	E	E	E	E	E	E	E	E	E	E	E	E	E	E	E	E	E
1500																											E	E	E	E	E
1600	E	E	E	E	E	E	E	E	E	E	E	E	E	E	E	E	E	E	E	E	E	E	E	E	E	E					
2300	D	D	D	D	D	D	D	D	D	D	D	D	D	D	D	D	D	D	D	D	D	D	D	D	D	D	D	D	D	D	D
	F	S	S	M	T	W	T	F	S	S	M	T	W	T	F	S	S	M	T	W	T	F	S	S	M	T	W	T	F	S	
NOV	1	2	3	4	5	6	7	8	9	10	11	12	13	14	15	16	17	18	19	20	21	22	23	24	25	26	27	28	29	30	
1100	E	E	E	E	E	E	E	E	E	E	E	E	E	E	E	E	E	E	E	E	E	E	E	E	E	E	E	E	E	E	
1500	E	E	E	E	E	E	E	E	E	E	E	E	E	E	E	E	E	E	E	E	E	E	E	E	E	E	E	E	E	E	
2300	D	D	D	D	D	D	D	D	D	D	D	D	D	D	D	D	D	D	D	D	D	D	D	D	D	D	D	D	D	D	
	S	M	T	W	T	F	S	S	M	T	W	T	F	S	S	M	T	W	T	F	S	S	M	T	W	T	F	S	S	M	T
DEC	1	2	3	4	5	6	7	8	9	10	11	12	13	14	15	16	17	18	19	20	21	22	23	24	25	26	27	28	29	30	31
1100	E	E	E	E	E	E	E	E	E	E	E	E	E	E	E	E	E	E	E	E	E	E	E				E	E	E	E	E
1500	E	E	E	E	E	E	E	E	E	E	E	E	E	E	E	E	E	E	E	E	E	E	E								
2300	D	D	D	D	D	D	D	D	D	D	D	D	D	D	D	D	D	D	D	D	D	D	D			D	D	D	D	D	D

50% Discounts for Caravans and Trailers

Towed caravans and trailers may travel at a discount of 50% of the applicable tariff on ALL sailings.

The company reserves the right to alter sailing times without prior notice.

- **It is essential to arrive at the departure dock not less than 45 minutes before all scheduled sailing times.**
- **Ships may sail up to 15 minutes before scheduled departure times.**

Table 2.6

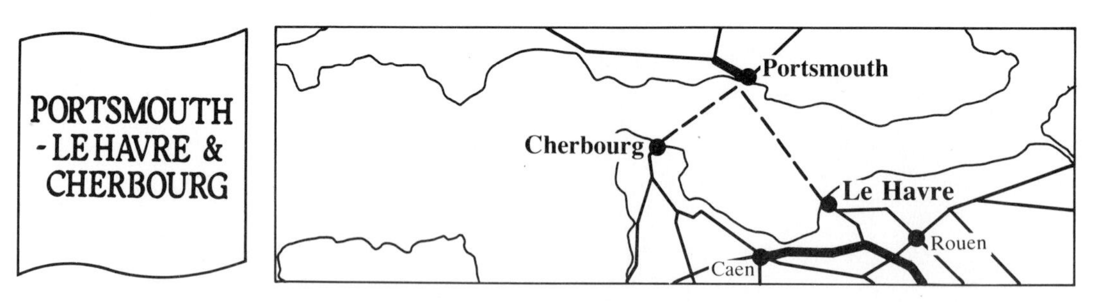

All fares are for single journeys unless otherwise stated *Port taxes included*

Travelling with a Vehicle	Tariff E £	Tariff D £	Tariff C £	Tariff B £	Cherbourg only Tariff A £
Drivers and Vehicle Passengers					
Adults	**18.00**	**18.00**	**18.00**	**18.00**	**18.00**
Children (4 and under 14 years. Under 4 free.)	**9.00**	**9.00**	**9.00**	**9.00**	**9.00**
Vehicles					
Cars, motor caravans, minibuses, vans (non-commercial use only) and motorcycle combinations.					
Overall length not exceeding 4.00m	**21.00**	**32.00**	**43.00**	**54.00**	**65.00**
4.50m	**21.00**	**41.00**	**54.00**	**65.00**	**76.00**
5.50m	**21.00**	**45.00**	**59.00**	**70.00**	**80.00**
over 5.50m, per additional metre or part thereof	**9.00**	**10.00**	**11.00**	**12.00**	**14.00**
Towed caravans and trailers (non-commercial use only).					
Overall length not exceeding 4.00m	**26.00**	**26.00**	**38.00**	**49.00**	**62.00**
5.50m	**26.00**	**38.00**	**55.00**	**74.00**	**85.00**
over 5.50m, per additional metre or part thereof	**15.00**	**15.00**	**15.00**	**15.00**	**15.00**
★★★ **The caravan fares listed are discountable by 50% on most sailings. See timetables for exceptions.**					
Motorcycles, scooters and mopeds	**9.00**	**10.00**	**11.00**	**12.00**	**13.00**
Foot Passengers and Cyclists					
Adults	**21.00**	**21.00**	**21.00**	**21.00**	**21.00**
Children (4 and under 14 years. Under 4 free.)	**10.50**	**10.50**	**10.50**	**10.50**	**10.50**
Cycles	**Free**	**Free**	**Free**	**Free**	**Free**

Reserved Accommodation

Night Sailings (2200-2330 hrs.)		**Day Sailings (0230-1930 hrs.)**	
4 berth cabin with shower and toilet per berth	**8.00**	4 berth cabin with shower and toilet	**11.50**
4 berth cabin, per berth	**7.00**	4 berth cabin	**8.50**
2 berth cabin with shower and toilet per berth	**12.50***	2 berth cabin with shower and toilet	**10.50***
2 berth cabin, per berth	**9.50**	2 berth cabin	**5.00**
Couchette, per berth	**4.00***	Club class seats	**2.00***
Club class seats (with rugs)	**4.00***	**Subject to availability.*	
Reclining seats (with rugs)	**2.00**		

Table 2.7

● Problems

Obtain current ferry brochures from a travel agent, and use them to find the travel cost for the following journeys.

1. Bristol–Paris via Portsmouth and Cherbourg with Townsend–Thoresen; sailing at lunch time on a Saturday in mid-July; 2 adults; BL Allegro.

2. Manchester–Bordeaux via Weymouth and Cherbourg with Sealink Ferries; sailing on a Saturday in August; 2 adult, 2 children (ages 6 and 9 years): Vauxhall Cavalier towing a caravan of length 4.3 m; 4 berths reserved on ferry. (If you find difficulty with this, then try the exercise on page 00.)

3. (i) Bristol–Nice via Dover and Calais with Hoverspeed; flight departure 1230 hours on a Monday in August; 4 adults; Ford Escort towing a trailer of length 2.3 m; overnight stop in France (bed and breakfast £4 per adult); toll charges £50; insurance cover £60.

(ii) What would be the difference in cost if the sailing were by
(a) Sealink,
(b) Townsend–Thoresen,
Compare the crossing times.

● **Related problems** Univers bold

Find the costs for the following.

A. Bristol–Paris via Dover and Calais with Townsend–Thoresen; sailing at lunch-time on a Saturday, in July; 2 adults; BL Allegro.

B. London–Paris via Dover and Calais with Townsend–Thoresen; sailing at lunch-time on a Saturday in July; 2 adults, 2 children (ages 10, 12 years); Ford Cortina.

C. Manchester–Bordeaux via Dover and Calais with Townsend–Thoresen, sailing on the first Saturday in August; 2 adults, 2 children (ages 6 and 9 years); Vauxhall Cavalier towing a caravan of length 4.3 m; overnight stops in France (bed and breakfast £5 per adult, £4 per child).

EXERCISE

All necessary information is given in Tables 2.2–2.7.

Manchester–Paris via Portsmouth and Le Havre with Townsend–Thoresen sailing at 23.30 hrs on Saturday 13th July; 2 adults, 2 children (ages 6 and 9 years); Vauxhall Cavalier towing a caravan of length 4.3 m; 4 berths reserved.

Distances (miles)

Manchester–Portsmouth ______

Le Havre–Paris ______

Total distance ______

Petrol costs

Consumption for Cavalier = ____ –8 = ____

Amount used $= \frac{342}{24}$ = ____ gal (to one decimal place)

Cost of petrol = 14.3 × £2 = £____

Ferry costs

Tariff for 2330 hrs crossing with Townsend–Thoresen Ferries on Saturday 13th July is letter ____

Vehicle, driver and 1 adult
(car not exceeding 4.5 m) = £ ____

2 children (4 and under 14 years) 2 x £ ____ = £ ____

Caravan (over 3 m) = £ ____

4-berth standard cabin 4 x £ ____ = £ ____

Total cost = £ ____

2.3. British Rail timetables

For this section you may need a copy of the current British Rail timetable.

British Rail timetables contain a great deal of information. Much of this information is in the form of symbols, letters, and numbers which need careful interpretation if details of journeys are to be obtained accurately. Explanation of the meaning of many symbols is given at the beginning of the book of timetables, and also at the bottom or side of many of the timetables.

We will now give an example of times of a journey from Nottingham to Bristol Temple Meads. This journey requires the use of more than one timetable, so it has been split up into parts to make it easier to understand.

DERBY TO BRISTOL TEMPLE MEADS

By looking up Derby in the pink pages of the 1985/86 timetable we see that Derby–Bristol times are contained in Table 56. Before using the table you must decide on what day you are travelling since Saturdays and Sundays often have different timetables from weekdays. You must also decide whether you want to travel from Bristol to Derby or Derby to Bristol. You are now ready to look at the specific part of the table relevant to your journey.

Suppose you wish to travel from Derby to Bristol on a weekday. Here is a copy of part of Table 56.

Table 56

Mondays to Fridays throughout, also Saturdays from 5 October

Derby → Birmingham, Cardiff and Bristol

Miles	Miles																								
			◇ MO 125	◇ MO	BHX	◇ A MO	BHX	BHX 2	SX		BHX	BHX C	BHX ◇ 125 Ø	◇ 2	◇ D	◇ 125 ☕	BHX ◇		◇	BHX 2 E	BHX 2 G	2		◇ 125 Ø	◇
0	—	**Derby** d													0602	0644	0717				0743			0829	0849
11	—	Burton-on-Trent d													0616	0655	0730				0800			0840	0903
24	—	Tamworth d													0629		0743				0816				0916
25¾	—	Wilnecote d													0633		0747				0820				
33	—	Water Orton 18 d													0646		0800								
41	—	**Birmingham New Street** 18 a													0701	0727	0815				0842			0912	0941
—	—	d	0011	0047		0535								0700		0732								0917	
67	—	Worcester Shrub Hill d									0700									0823			0920		
89	—	Cheltenham Spa d	0054	0130	0525		0617				0722		0738	0748f		0812			0830	0851			0945a	0957	
95½	0	Gloucester a	0106	0143	0535		0627				0730		0745						0838	0901				1007	
—	—	d	0111	0154							0741	0737										0910		1012	
—	19½	Lydney d										0757										0930			
—	27½	Chepstow d						0655	0735			0807										0940			
—	33¾	Caldicot d						0703	0743			0815										0948			
—	34¾	Severn Tunnel Junction 131 a						0706	0746			0818										0951			
—	44¼	Newport 131 a				0811b		0718	0757		0829	0829				0933g						1002		1122g	
—	56¼	**Cardiff Central** 131 a				0828b		0750c	0815		0847e	0847				0949g						1020		1139g	
128½	—	Bristol Parkway 131 a	0140			0700					0816			0832s		0848								1041	
134¾	—	**Bristol Temple Meads** 131 a	0154	0240		0715					0829			0849		0900								1055	

		2 J	◇ 125 Ø			2	K			◇ ☕	◇		◇ L FO ☕	◇ N MO ☕	◇ P ☕	2 Q			◇ 125 ☕	125 Ø	◇ Ø	2		◇ U Ø ☕	
Derby d		0914	0934							1020			1110	1110	1110	1126					1149			1238	
Burton-on-Trent d		0928								1034						1140					1203				
Tamworth d		0944														1155									
Wilnecote d		0948														1200									
Water Orton 18 d																									
Birmingham New Street 18 a		1009	1015							1111			1150	1157	1158	1221					1241			1325	
d			1019							1115	1130		1202	1202	1202					1225				1330	
Worcester Shrub Hill d							1043	1124																	
Cheltenham Spa d					1030			1149a		1200									1302	1310			1330		
Gloucester a					1039					1210									1309	1320			1340		
d										1210		1222						1235		1325					
Lydney d																		1255							
Chepstow d				1025		1125												1305				1335			
Caldicot d				1033		1133												1313				1343			
Severn Tunnel Junction 131 a				1036		1136												1316				1346			
Newport 131 a			1227b	1048		1147	1233	1256		1256		1347g						1328		1439g		1358			
Cardiff Central 131 a			1245b	1128	1315e	1205	1250	1315h		1315		1403g						1355		1455g		1455			
Bristol Parkway 131 a			1129									1256								1354				1541b	
Bristol Temple Meads 131 a			1143								1259	1311	1331	1331	1331					1408				1500	

	K SX		2	◇ 125 Ø	2 V				2 Y	◇ 125 Ø	◇ ☕	2 Z	AA		2		K	◇ ☕	◇ 125 ☕	K	2 BB	◇ ☕	2 CC		
Derby d					1326					1448		1513						1550	1616				1700		
Burton-on-Trent d					1340							1527						1604					1714		
Tamworth d					1355							1543											1730		
Wilnecote d					1400							1547											1734		
Water Orton 18 d																									
Birmingham New Street 18 a					1423					1529		1608						1640	1657				1755		
d				1358						1532	1544								1702			1733			
Worcester Shrub Hill d	1232	1405															1626			1735	1728				
Cheltenham Spa d		1432		1441			1530			1614	1631			1730					1742		1755	1821			1834
Gloucester a		1442		1451			1540			1624				1739					1805k		1805	1833			1843
d				1456		1500				1624			1650		1746							1833			
Lydney d						1520									1806										
Chepstow d			1445			1530			1600						1816										
Caldicot d			1453			1538			1608						1824										
Severn Tunnel Junction 131 a			1456			1541			1611						1827										
Newport 131 a	1444		1508			1553			1622	1705			1830g		1839		1843		1849n	1917		1927s			2038p
Cardiff Central 131 a	1530		1530			1640			1640	1722			1846g		1905j		1905		1905n	1934		1953			2055p
Bristol Parkway 131 a				1526							1712		1726						1815						
Bristol Temple Meads 131 a				1540							1728		1739						1829						

For general notes see pages 2 - 4

Voir pages 5 - 7 de l'indicateur pour les renseignements généraux et l'explication des signes

Allgemeine Bemerkungen und Zeichenerklärung siehe Seiten 8 - 10 des Kursbuches

For details of meals service available on trains marked Ø see train catering section at the front of this Timetable

- **A** 27 May to 9 September
- **C** From Swindon dep. 0638 (Table 126)
- **D** To Brighton arr. 1100 (Table 51)
- **E** From Birmingham New Street dep. 0732 (Table 67)
- **G** From Nottingham dep. 0708 (Table 80)
- **J** From Lincoln Central dep. 0732 (Table 30)
- **K** Via Hereford (Tables 126 and 87)
- **L** 13 and 20 December, 3 January, 4 April and 2 May
- **N** 16 December
- **P** Mondays to Fridays until 27 September, and Saturdays 14 December and 3 and 10 May, and Tuesdays, Wednesdays, Thursdays and Saturdays 17 to 21 December, and Thursday 2 January, and Wednesday and Thursday 26 and 27 March, and Tuesday, Wednesday and Thursday 1,2 and 3 April
- **Q** From Cleethorpes dep. 0823 (Table 30)
- **U** ☕ Saturdays
- **V** From Nottingham dep. 1253 (Table 80)
- **Y** Second Class only Saturdays
- **Z** From Nottingham dep. 1440 (Table 80)
- **AA** To Weston-super-Mare arr. 1814 (Table 134)
- **BB** From Birmingham New Street dep. 1632 (Table 67)
- **CC** From Nottingham dep. 1623 (Table 80)
- **b** Change at Bristol Temple Meads
- **c** Saturdays arr. 0828
- **e** Change at Gloucester and Newport
- **f** Arr. 0744
- **g** Change at Bristol Parkway
- **h** Change at Cheltenham Spa and Newport
- **j** Saturdays arr. 1910
- **k** Second Class only
- **n** Change at Bristol Parkway. Saturdays arr. Newport 1854, Cardiff Central 1910
- **p** Change at Gloucester and Bristol Parkway

Enterprising mathematics

All the trains are straight through (no change is required), if the arrival time is in bold print. Italic print denotes change required.

Suppose you didn't mind changing trains, and left Derby at 14.48. There is a through train to Birmingham, arriving at 15.29. Changing there, you can depart from Birmingham at 15.44 and arrive at Bristol at 17.28.

NOTTINGHAM–DERBY

A single table for Bristol–Nottingham was not listed, that is why we looked at Bristol–Derby firstly. We now look for train times between Nottingham and Derby; using Table 80. All the trains are straight through.

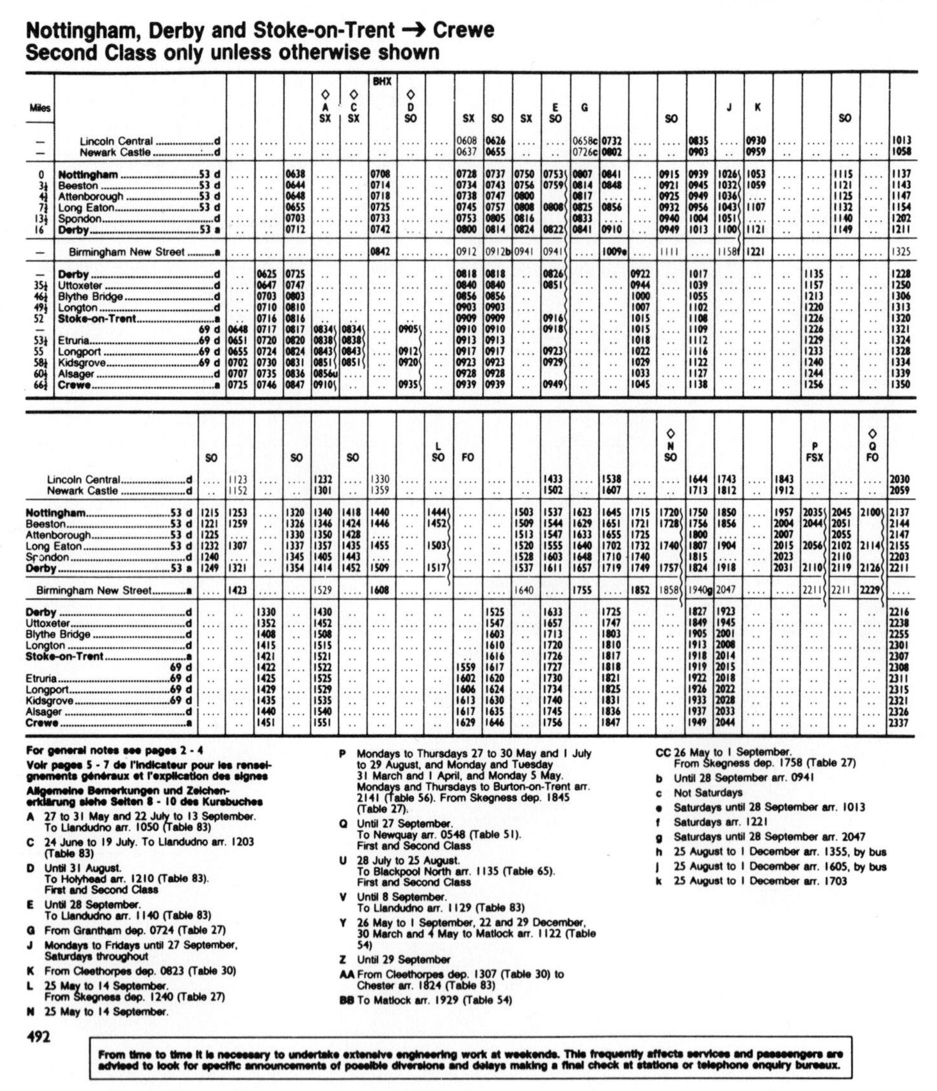

Table 80 **Mondays to Saturdays**

Nottingham, Derby and Stoke-on-Trent → Crewe
Second Class only unless otherwise shown

Miles					◇ A SX	◇ C SX	BHX	◇ D SO		SX	SO	SX	E SO	G			SO		J	K			SO		
—	Lincoln Central d									0608	0626			0658c	0732			0835		0930					1013
—	Newark Castle d									0637	0655			0726c	0802			0903		0959					1058
0	**Nottingham** 53 d			0638			0708			0728	0737	0750	0753	0807	0841		0915	0939	1026	1053			1115		1137
3¼	Beeston 53 d			0644			0714			0734	0743	0756	0759	0814	0848		0921	0945	1032	1059			1121		1143
4¾	Attenborough 53 d			0648			0718			0738	0747	0800		0817			0925	0949	1036				1125		1147
7¾	Long Eaton 53 d			0655			0725			0745	0757	0808	0808	0825	0856		0932	0956	1043	1107			1132		1154
13¼	Spondon d			0703			0733			0753	0805	0816		0833			0940	1004	1051				1140		1202
16	**Derby** 53 a			0712			0742			0800	0814	0824	0822	0841	0910		0949	1013	1100	1121			1149		1211
—	Birmingham New Street a						0842			0912	0912b	0941	0941		1009e		1111		1158f	1221					1325
—	**Derby** d		0625	0725						0818	0818		0826			0922		1017				1135			1228
35½	Uttoxeter d		0647	0747						0840	0840		0851			0944		1039				1157			1250
46½	Blythe Bridge d		0703	0803						0856	0856					1000		1055				1213			1306
49½	Longton d		0710	0810						0903	0903					1007		1102				1220			1313
52	**Stoke-on-Trent** a		0716	0816						0909	0909		0916			1015		1108				1226			1320
—	69 d	0648	0717	0817	0834	0834		0905		0910	0910		0918			1015		1109				1226			1321
53¼	Etruria 69 d	0651	0720	0820	0838	0838				0913	0913					1018		1112				1229			1324
55	Longport 69 d	0655	0724	0824	0843	0843		0912		0917	0917		0923			1022		1116				1233			1328
58¼	Kidsgrove 69 d	0702	0730	0831	0851	0851		0920		0923	0923		0929			1029		1122				1240			1334
60¼	Alsager d	0707	0735	0836	0856u					0928	0928					1033		1127				1244			1339
66¾	**Crewe** a	0725	0746	0847	0910			0935		0939	0939		0949			1045		1138				1256			1350

	SO			SO		SO			L SO	FO							◇ N SO					P FSX		◇ Q FO	
Lincoln Central d		1123			1232		1330						1433		1538			1644	1743		1843				2030
Newark Castle d		1152			1301		1359						1502		1607			1713	1812		1912				2059
Nottingham 53 d	1215	1253		1320	1340	1418	1440		1444			1503	1537	1623	1645	1715	1720	1750	1850		1957	2035	2045	2100	2137
Beeston 53 d	1221	1259		1326	1346	1424	1446		1452			1509	1544	1629	1651	1721	1728	1756	1856		2004	2044	2051		2144
Attenborough 53 d	1225			1330	1350	1428						1513	1547	1633	1655	1725		1800			2007		2055		2147
Long Eaton 53 d	1232	1307		1337	1357	1435	1455		1503			1520	1555	1640	1702	1732	1740	1807	1904		2015	2056	2102	2114	2155
Spondon d	1240			1345	1405	1443						1528	1603	1648	1710	1740		1815			2023		2110		2203
Derby 53 a	1249	1321		1354	1414	1452	1509		1517			1537	1611	1657	1719	1749	1757	1824	1918		2031	2110	2119	2126	2211
Birmingham New Street a		1423			1529		1608					1640		1755		1852	1858	1940g	2047			2211	2211	2229	
Derby d			1330		1430						1525		1633		1725			1827	1923						2216
Uttoxeter d			1352		1452						1547		1657		1747			1849	1945						2238
Blythe Bridge d			1408		1508						1603		1713		1803			1905	2001						2255
Longton d			1415		1515						1610		1720		1810			1913	2008						2301
Stoke-on-Trent a			1421		1521						1616		1726		1817			1918	2014						2307
69 d			1422		1522					1559	1617		1727		1818			1919	2015						2308
Etruria 69 d			1425		1525					1602	1620		1730		1821			1922	2018						2311
Longport 69 d			1429		1529					1606	1624		1734		1825			1926	2022						2315
Kidsgrove 69 d			1435		1535					1613	1630		1740		1831			1933	2028						2321
Alsager d			1440		1540					1617	1635		1745		1836			1937	2033						2326
Crewe a			1451		1551					1629	1646		1756		1847			1949	2044						2337

For general notes see pages 2 - 4

Voir pages 5 - 7 de l'indicateur pour les renseignements généraux et l'explication des signes

Allgemeine Bemerkungen und Zeichenerklärung siehe Seiten 8 - 10 des Kursbuches

A 27 to 31 May and 22 July to 13 September. To Llandudno arr. 1050 (Table 83)

C 24 June to 19 July. To Llandudno arr. 1203 (Table 83)

D Until 31 August. To Holyhead arr. 1210 (Table 83). First and Second Class

E Until 28 September. To Llandudno arr. 1140 (Table 83)

G From Grantham dep. 0724 (Table 27)

J Mondays to Fridays until 27 September, Saturdays throughout

K From Cleethorpes dep. 0823 (Table 30)

L 25 May to 14 September. From Skegness dep. 1240 (Table 27)

N 25 May to 14 September.

P Mondays to Thursdays 27 to 30 May and 1 July to 29 August, and Monday and Tuesday 31 March and 1 April, and Monday 5 May. Mondays and Thursdays to Burton-on-Trent arr. 2141 (Table 56). From Skegness dep. 1845 (Table 27).

Q Until 27 September. To Newquay arr. 0548 (Table 51). First and Second Class

U 28 July to 25 August. To Blackpool North arr. 1135 (Table 65). First and Second Class

V Until 8 September. To Llandudno arr. 1129 (Table 83)

Y 26 May to 1 September, 22 and 29 December, 30 March and 4 May to Matlock arr. 1122 (Table 54)

Z Until 29 September

AA From Cleethorpes dep. 1307 (Table 30) to Chester arr. 1824 (Table 83)

BB To Matlock arr. 1929 (Table 54)

CC 26 May to 1 September. From Skegness dep. 1758 (Table 27)

b Until 28 September arr. 0941

c Not Saturdays

e Saturdays until 28 September arr. 1013

f Saturdays arr. 1221

g Saturdays until 28 September arr. 2047

h 25 August to 1 December arr. 1355, by bus

j 25 August to 1 December arr. 1605, by bus

k 25 August to 1 December arr. 1703

492

From time to time it is necessary to undertake extensive engineering work at weekends. This frequently affects services and passengers are advised to look for specific announcements of possible diversions and delays making a final check at stations or telephone enquiry bureaux.

Suppose you wished to catch the 08.29 train from Derby to Bristol. Looking at Table 80, you would have to leave Nottingham at 07.50, arrive Derby at 08.24. After a 5 minute wait you can catch the 08.29 as required. Of course you may wish to give yourself more time at Derby so you could catch the 07.37 arriving at 08.14.

● Problems

1. If you leave Nottingham after 09.00 on a weekday, what is the earliest time that you can arrive at Bristol Temple Meads?

2. If you wish to arrive at Bristol Temple Meads before 18.00, what is the latest time you can leave Nottingham?

3. Work out details of a journey from Darlington to Leeds leaving Darlington after 11.00 on a weekday. (You will need a copy of the current British Rail timetable.)

2.4. Making the most of your Railrover

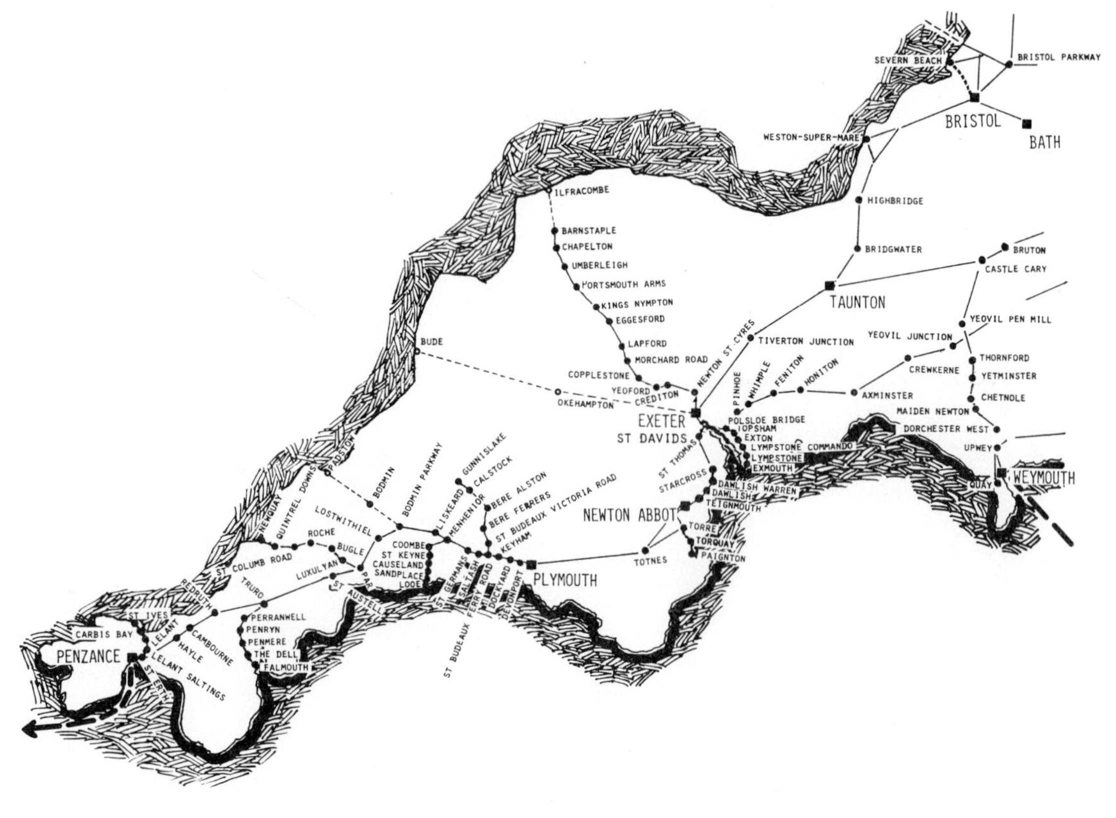

Fig. 2.2

For this section you will require a copy of the current British Rail timetable.

A 'West of England' Railrover gives you unlimited travel for a week on all routes in the southwest. These are shown on the route map in Fig. 2.2.

You use the Railrover for two sightseeing trips on Saturday and Sunday. For the remaining 5 days you have challenged 4 friends to a competition in which each of you uses the Railrover for one day and travels as far as possible, starting and finishing at Salisbury. The winner is the one who travels furthest.

How far can you travel in one day?

The distance between stations is given on the first page of each table, down the left-hand side.

2.5. Youth hostelling

Carrock Fell in Cumbria - our newest hostel.

Youth hostelling is a cheap way to explore the countryside (and some cities too) provided you are prepared to move 'under your own steam'. A holiday in a particular area can be planned by choosing hostels within a day's travel of each other (making sure the hostel is open on the day you plan to stay there!). A tour of the Brecon Beacons and mid-Wales might look something like this

Ty'n-y-Caeau
Ystradfellte
Dolgoch
Knighton
Malvern Hills

But how cheap is cheap? Charges at hostels depend upon the grade of the hostel, VAT or non-VAT hostel, and the age of the member. The above tour for a junior member (16–21 years) would cost

Ty'n-y-Caeau	£2.15	(Standard)
Ystradfellte	£1.65	(Simple, non-VAT)
Dolgoch	£1.65	(Simple, non-VAT)
Knighton	£1.90	(Standard, non-VAT)
Malvern Hills	£2.15	(Standard)
	£9.50	

Meals and sleeping bag hire charges (if necessary) are additional to this.

Table 2.8. Hostel charges from March 1st, 1983

Grade Hostels are graded according to the facilities they provide.
VAT At most hostels Value Added Tax is payable on all hostel charges except lunch packets. Some of the smaller or lesser used hostels are exempt from VAT. They are marked "Non VAT" in the hostel details.

Overnight fees

VAT hostels			
Grade	*Young*	*Junior*	*Senior*
Simple	£1.50	£1.85	£2.20
Standard	£1.80	£2.15	£2.75
Superior	£2.05	£2.50	£3.10
Special	£2.25	£2.85	£3.50
London			
Holland House	£3.15	£3.60	£4.30
Earl's Ct., Carter Lane, Hampstead	£2.75	£3.25	£3.95
Highgate	£2.30	£2.95	£3.60
Non VAT hostels			
Grade	*Young*	*Junior*	*Senior*
Simple	£1.30	£1.65	£1.95
Standard	£1.60	£1.90	£2.40
Superior	£1.80	£2.20	£2.70

Meal prices

Evening meal	£1.60 (non VAT £1.40)
Breakfast	£1.30 (non VAT £1.15)
Lunch packet	85p (VAT does not apply)

Sleeping bag hire 65p (non VAT 60p)

Field study charges per day

Grade	*Pre-Sixth form*	*Sixth form and tertiary education*
Standard	£6.70	£7.00
Superior	£7.00	£7.50

Car, coach and motor-cycle parking free.

The YHA does not accept liability for loss or damage to vehicles or their contents while parked at hostels.
Camping At those hostels where camping is allowed, the charge per person is half the senior overnight fee for that hostel, regardless of the age of the member.

All charges stated herein are those current at the time of printing, but they are liable to change during a year. It is a condition of booking that, notwithstanding the payment of charges printed herein and the confirmation of the booking, all persons are obliged to pay the charges current at the time of taking up their booking. When there has been any increase any additional charges must be paid to the Warden of the hostel at the time of arrival

Table Classification of Youth Hostels

Bakewell	Standard VAT
Boggle Hole	Superior VAT
Hagg Farm	Standard VAT
Helmsley	Standard VAT
Lockton	Simple non-VAT
Matlock Bath	Standard VAT
Osmotherley	Superior VAT
Whitby	Standard VAT
Windgather Cottage	Simple non-VAT

● Problems

1. A tour of the Peak district would be: Matlock Bay, Windgather Cottage, Hagg Farm, and Bakewell. Calculate the hostel charges for a young or junior member.

2. The Yorkshire Moors are well worth a visit. Osmotherley, Helmsley, Lockton, Boggle Hole, and Whitby could form the basis of a tour. What is the cost if each night you need to hire a sleeping bag?

3. Decide upon a tour you would like to do and work out the cost. How much extra would you spend if you bought an evening meal and packed lunch each day of your tour?

Youth Hostel Charges

These are given in Table 2.8.

3 Parties

This chapter can be dealt with as a whole or individual sections can be taken in any order. Each stands alone but they have a common theme built around an eighteenth birthday party.

A project on this theme could start with a class discussion. Topics which might arise are how to pay for food and drink, restrict admission to those invited, and how to provide music.

The question of how to organize and pay for the whole event should make pupils think about how much they would be willing to spend on their own party.

Solutions: Some questions are open-ended. The answers to calculations are given.

3.1. How much from a bottle?

We expect beer to be served in pints or half pints so it is easy to work out how much beer or cider you should get for a party.

It's not so easy with wine. There are many different sizes of wine glass but no standard measure for wine by the glass. You know how many half pints you can get from a four-pint can of beer, but have you any idea how many glasses of wine you can get from a 1-litre bottle?

You might like to pour out what seems the right amount, and measure what this is. It is likely to be between $2\frac{1}{2}$ and 4 fl oz. How many glasses would you get from $1\frac{1}{2}$ litres or from a standard 75-cl bottle?

$$\begin{aligned} 1 \text{ litre} &= 35 \text{ fl oz} \\ &= 100 \text{ cl} \\ &= 1\tfrac{3}{4} \text{ pints approx.} \end{aligned}$$

If you allow 3 fl oz for a glass, 1 litre will give about 12 glasses so $1\frac{1}{2}$ litres will give 18. A standard bottle contains half that quantity and so gives 9 glasses.

What is the most economical way to buy for a party if you mean to get 3 litres and find that

(i) 3-litre box costs £6.79;
(ii) $1\frac{1}{2}$-litre bottle costs £3.19;
(iii) 75-cl bottle costs £2.04?

Working out the three alternatives gives:

(i) 3-litre box costs £6.79;
(ii) 2 × 1½-litre bottle costs 2 × £3.19 = £6.38;
(iii) 4 × 75-cl bottle costs 4 × £2.04 = £8.16.

If the wines are the same, it looks as though the 1½-litre bottle is the best buy for a party. Single 75-cl bottles look expensive but prices can vary from place to place.

● Problems

1. How many 4 fl oz glasses can you get from a 1½-litre bottle?

2. How many glasses of 2½ fl oz can you get from 1½-litres?

3. If you're having a party and want to allow 30 people 3 glasses of 4 fl oz each, how many 1½-litre bottles would you have to buy?

4. For 20 people how many litre bottles must you buy to give each person 4 glasses of 5 fl oz each?

5. How much beer and lemonade must you buy to give 30 half-pint glasses of shandy if it is to contain 1/3 beer and 2/3 lemonade?

6. If you have £20 to spend on wine for a party, how many 1½-litre bottles can you buy? What total volume in fl oz does this give?

7. A party can of 7 pints of beer costs £3.39. Is this a better or worse bargain than the 4-pint can at £2.30?

8. A 34-pint barrel of bitter can be bought for £16.70. Is this a better buy per pint than the previous types?

9. Now you can make a list for a party of your own. If you tell everyone to bring a bottle, what do you expect them to bring? How much are you willing to provide? Write down what you intend to allow for each person and work out what the total cost would be.

3.2. Can you cook?

If you were having a party, how would you decide whether to buy ready-made cakes or make your own (or get your Mum busy)? One thing to consider is cost. For example, at the local shop cheesecake costs 45p a portion. So, if you were expecting 30 people you would have to pay out 30 X 45p = £13.50.

The ingredients shown below are for a chocolate cheesecake serving 6–8 people. The costs of the individual ingredients are found in Table 3.1.

Table 3.1

Groceries		Cost (pence)	Fresh food		Cost (pence)
Flour	1.5 kg	39	Eggs	6 size 2	47
Cornflour	375 g	39	Eggs	6 size 4	41
Caster sugar	1 kg	42	Butter	250 g	45
Icing sugar	500 g	33	Milk	1 pt	19
Plain chocolate	150 g	66	Cottage cheese	225 g	44
Pineapple slices	per can	22	Full fat cheese	225 g	56
Bourbon biscuits	per pkt	23	Single cream	½ pt (284 ml)	54
Sponge fingers	per pkt	26	Double cream	½ pt (284 ml)	11
Cheesecake	per portion	45			

Off licence		Cost
Brandy	50 cl	£4.25

Ingredients	*Approximate cost (£)*
1 pkt bourbon biscuits	0.23
150 g butter	0.27
225 g cottage cheese	0.44
225 g full fat soft cheese	0.56
50 g caster sugar	0.02
juice of ½ lemon	0.06
½ pt single cream	0.54
1 tablespoon powdered gelatine	very small so ignored
75 g plain dessert chocolate	0.33
Total	£2.45

Each cake costs approximately £2.50. You want to serve 30 so need to make 4 cakes. This would cost about 4 × £2.50 = £10 and you should allow for decoration etc. say £1, £11 altogether. Making your own cakes would save you about

$$£13.50 - £11.00 = £2.50.$$

The difference in cost is probably small enough not to matter much when other factors are taken into account.

● Problems

When working out the prices in the following problems, you will need to refer to the prices at the end. Note, though, that you cannot always buy exactly the amount you want. For example, the chocolate cake needed 150 g of butter, but you would have to buy 250 g.

250 g butter costs 45 p;
50 g would cost 45 p ÷ 5 = 9 p;
150 g costs 9 p × 3 = 27 p.

1. How much would it cost for one portion of the following cake?

Chocolate gateaux (4–6 people)

Ingredients

2 boxes sponge fingers
¼ pt (150 ml) milk
2 tablespoons (40 ml) brandy or rum
4 oz (100 g) plain chocolate
1 teaspoon instant coffee powder
6 oz (175 g) butter
6 oz icing sugar
2 egg yolks
¼ pt (150 ml) cream

2. How much would it cost to make up a packet cheesecake mix costing 63 p which needs 1/3 pint milk and 50 g butter?

3. Find a cake recipe that appeals to you and work out the cost of providing one portion each for 25 people.

4. What would your bill come to if you went shopping for the ingredients for

   (a) 1 pineapple gateau;
   (b) 4 pineapple gateaux?

(Assume that the smallest amounts you can buy are those given in the price list in Table 3.1.)

Ingredients for pineapple gateau

4 oz (100 g) butter
3 oz (75 g) plain flour
4 large eggs
4 oz (100 g) caster sugar
1 oz (25 g) cornflour
½ pt double cream
can pineapple slices
2 oz (50 g) icing sugar

● Related problems

A. For one cheesecake, using the ingredients given, what would you have to buy that you would not use?

B. How much would it cost to make 3 pineapple gateaux if you only count the cost of the actual ingredients?

C. Look at the ingredients for the chocolate gateau. It gives

   4 oz (100 g) plain chocolate
   6 oz (175 g) icing sugar

   (a) Are these conversions the same?
   (b) Can you explain this?

D. What other factors, besides cost, would you think about before deciding whether to make your own gateaux?

3.3. Cost of food

Friends may bring drinks to a party but they're not so likely to bring food. What would you have to eat? How much would it cost you to provide food for a party for 30 using the price list in Table 3.2?

Table 3.2

Item	Cost	Item	Cost
36 sausage rolls	£1.44	4 pizzas	£1.80
10 individual pork pies	£1.36	1 lb cheese	£1.10
8 chicken drumsticks	£2.16	20 sausages (2½ lb)	£2.32
6 packs of crisps	£0.39	Chocolate cake	£2.50
18 chocolate eclairs	£1.98	Sliced loaf (25 slices)	£0.30
400 g nuts	£1.20	½ lb cold meat	£1.08
packet cheese biscuits	£0.42	250 g butter	£0.45

72 sausage rolls	2 × £1.44	= £2.88
30 pork pies	3 × £1.36	= £4.08
24 chicken drumsticks	3 × £2.16	= £6.48
4 packets cheese biscuits	4 × £0.42	= £1.68
2 loaves	2 × £0.30	= £0.60
500 g butter	2 × £0.45	= £0.90
1 lb cold meat	2 × £1.08	= £2.16
400 g nuts	1 × £1.20	= £1.20
Total		= £19.98

If you allow £20 for 30 people this will cost you about 70 p for each person, and allow a reasonable quantity of food.

● Problems

1. (a)

4 cans orange	£1.20
100 g coffee	£0.84
2 tubs icecream	£1.05
150 g biscuits	£0.19
4 cans lager	£1.50
1 l cider	£0.75

(b)

200 g coffee	£0.99
5 lb chips	£0.89
5 lb peas	£1.38
400 g dog food	£0.40
4 l cooking oil	£2.08
200 g jar coffee	£1.22

(c)

10 doughnuts	£1.70
500 g drinking chocolate	£1.37
4 lb chips	£0.88
2 lb cod	£1.76
2 l cooking oil	£1.19
750 g tin coffee	£6.29

(d)

3 kg dog food	£1.05
2 lb peas	£0.82
4 lb chips	£0.95
800 g cod	£1.59
1 kg drinking chocolate	£1.89
4 tubs ice cream	£1.55

Now use the prices given in Table 3.2 to work out the costs for the lists below. You can only buy in the sizes named, and cannot buy parts of packets.

2. Work out the cost of this shopping list to cater for 24 people

4 loaves
750 g butter
2 lb meat
12 pizzas
24 packets crisps
1 chocolate cake.

3. What would be the price of 5 lb sausages, 32 chicken drumsticks, and 12 packets of crisps? Would you have enough change from £15 to pay for 800 g nuts?

4. If you buy 6 french loaves at 46 p each, 4 lb cheese, and 750 g butter how many boxes of chocolate eclairs can you buy if you have £12.50 altogether to spend?

5. Find the price of 10 sausage rolls, 10 pork pies, 10 chocolate eclairs, and 10 packets of crisps if the items could all be bought separately at the same prices as before.

If you earn 70 p an hour for a Saturday job, how many complete hours would you have to work to pay for this?

6. If the quality is the same in both cases, work out which of these is the better buy.

(a) 50 fish fingers for £1.59
or 30 fish fingers for £1.29

(b) 200 g coffee for £1.29
or 750 g for £4.99.

7. Which is the cheapest type of teabag per 100?

440 type A cost £1.95;
275 type B cost £1.99;
200 type C cost £0.99.

Which type would you buy? Why would you choose these?

8. Is the heaviest the dearest per oz for the following?

24 × 2 oz beefburgers cost £3.48
24 × 1¾ oz beefburgers cost £2.83
12 × 3½ oz beefburgers cost £3.27
8 × 4 oz beefburgers cost £2.79
20 × 1¾ oz beefburgers cost £2.29.

Which is the cheapest per oz?

9.

Sausages	Price per pack	Number in 2½lb pack
Pork thick	£1.92	24
Pork thin	£2.07	48
Pork and beef	£1.47	24
Pork thick	£1.75	20
Pork and beef	£1.55	20
Pork thin	£1.85	40

Which is the cheapest per lb of these sausages?

For a party 2½ lb pack, which would give you most items for the money?

10. Now choose a menu for a party. You must decide how many people to invite, and how much you intend to spend on food.

Write down your shopping list and show the quantity you intend to buy of each item.

Find out what each item costs in the shops this week and make out a bill for what your list would cost. Round this off to a convenient total.

If the total is more than your budget allows, you will need to reduce the menu or invite fewer people. You will have to decide what to do about this before going any further.

Show what you intend to buy, and finish off with a total showing all you plan to spend on food.

3.4. Running a bar at a disco

If you run a disco for a large number of people, the easiest way of providing drinks is to get a licence and run your own bar. It can also be a good way to make money for your club or school. The problems are, how much of each sort of drink do you buy, and how much to you charge your customers?

If you buy too much drink, and have no arrangement with the shop to buy it back (sale-or-return), your profit will be reduced by what you do not sell. If your prices are too low, you won't make enough profit (though sales may be high); if prices are too high you still won't make a big profit, because you will sell less.

You have to make sensible estimates of quantities and prices, based on past experience.

BACKGROUND INFORMATION

Table 3.3 gives the price lists, and the number of glasses sold, at three different discos, each for 200 people.

The glass sizes at all the discos were

Beer	½ pint
Wine	1/8 litre (8 glasses from one litre)
Coke	1/6 litre (6 glasses from one litre).

Stock was bought on a sale-or-return basis where only whole bottles or cans could be returned. The cost of buying stock is shown in Table 3.4.

Table 3.3

Drink	Disco A		Disco B		Disco C	
	Price per glass (£)	Number of glasses sold	Price per glass (£)	Number of glasses sold	Price per glass (£)	Number of glasses sold
Beer	0.24	150	0.30	130	0.36	90
Wine	0.25	100	0.35	80	0.45	50
Coke	0.10	200	0.15	180	0.20	150

Table 3.4

Drink	Size of container bought	Cost (£)
Beer	7-pint can	3.00
Wine	2-litre bottle	3.50
Coke	2-litre bottle	0.60

WORKING OUT THE PROFIT

As an example, look at the beer sold at Disco B.

The takings were £0.30 × 130 = £39.00.
The 130 ½-pint glasses needed 65 pints of beer = 65/7 = 9.3 seven-pint cans.
Therefore 10 whole cans were needed, so total costs were £3 × 10 = £30.
Profit on beer = £39 – £30 = £9 for disco B.

● Problems

1. Work out the profit at Disco B for wine and for coke. (If you need more help, refer to the worksheet at the end of this section.) Now work out the total profit made at Disco B.

2. Work out the total profits made at Disco A and at Disco C. Which of the three price lists gave the biggest profit?

● Related problems

A. From the first row of Table 3.3 we can make a table showing the prices and sales of beer at the three discos.

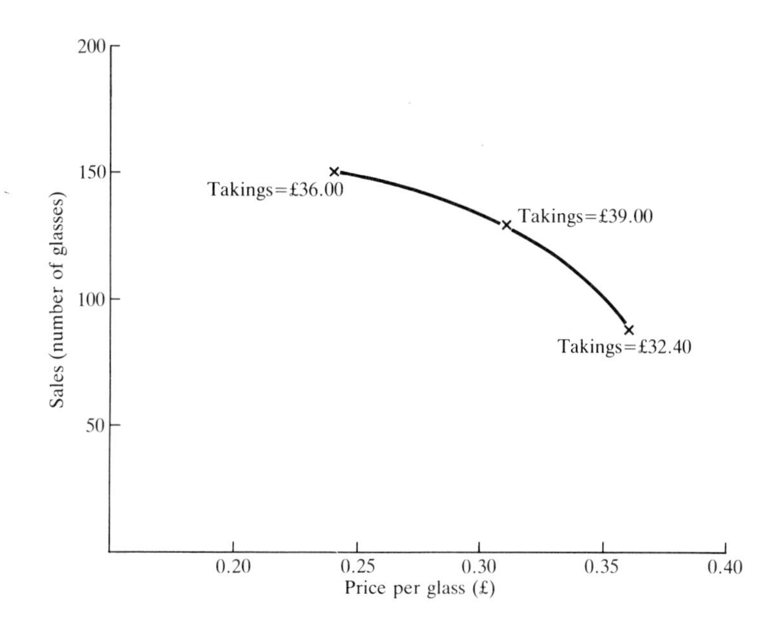

Fig. 3.1

	Price per glass (£)	Sales (number of glasses sold)
Disco A	0.24	150
Disco B	0.30	130
Disco C	0.36	90

Fig 3.1 is a graph plotted from this information. It shows how the sales of beer drop as the price goes up. Besides each point are the 'takings' in pounds for each combination of prices and sales.

Estimate from Fig. 3.1 the number of glasses sold for a price of 29 p (£0.29). Work out the takings for this number. In the same way, investigate other points on the graph to see which price gives the biggest takings.

B. Draw graphs similar to Fig. 3.1 for wine and coke. Investigate them in the same way.

● Worksheet

How many ½-pint beer glasses do you fill from a 7-pint can?
How many 1/8-litre wine glasses do you fill from a 2-litre bottle?
How many 1/6 litre coke glasses do you fill from a 2-litre bottle?

For Disco B, complete the following tables and questions.

Drink	Price per glass (£)	Number of glasses sold	Takings (£)
Beer	0.30	130	39.00*
Wine	0.35	80	
Coke	0.15	180	

* £0.30 × 130 = £39.00.

What are the total takings for Disco B?

Drink	Number of glasses sold	Number of cans or bottles needed	Cost per bottle or can (£)	Total cost of particular drink (£)
Beer	130	10*	3.00	30.00†
Wine	80			
Coke	180			

* 130 ÷ 14 = 9.3; round up to 10 whole cans.
† £3.00 × 10 = £30.00.

What is the total cost of all the drinks?

If

$$\text{Profit} = \text{Total takings} - \text{Total cost},$$

what is the profit on drinks for Disco B?

3.5. Don't drink and drive

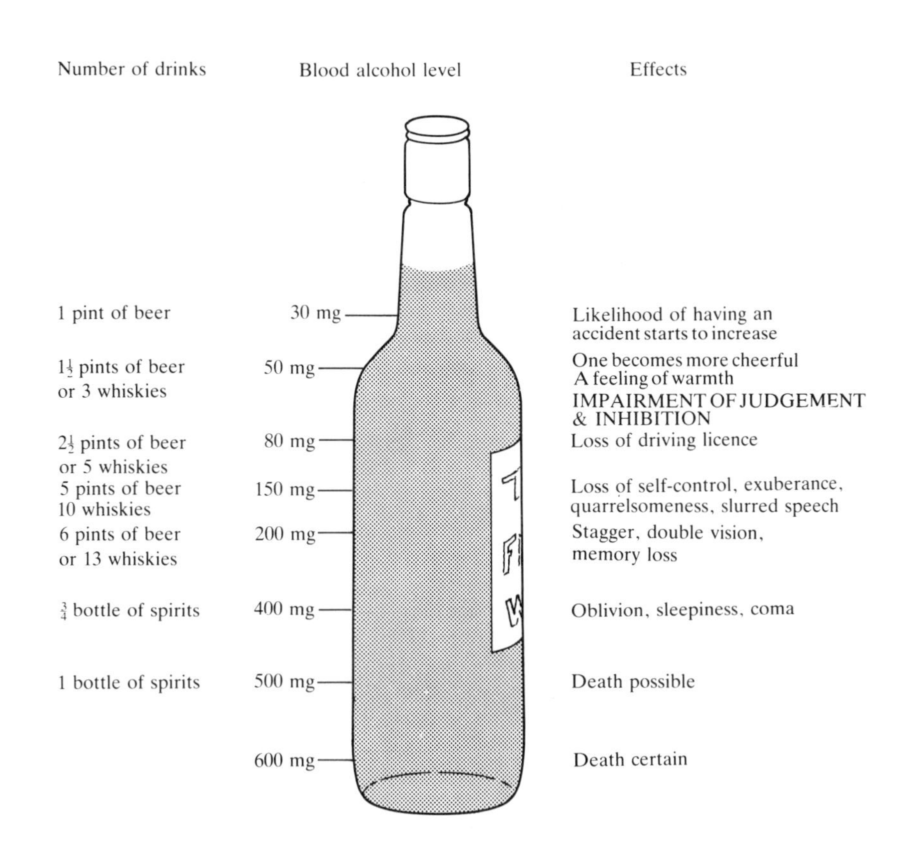

Number of drinks
Blood alcohol level
Effects
1 pint of beer
30 mg
Likelihood of having an accident starts to increase
1½ pints of beer or 3 whiskies
50 mg
One becomes more cheerful
A feeling of warmth
IMPAIRMENT OF JUDGEMENT & INHIBITION
2½ pints of beer or 5 whiskies
80 mg
Loss of driving licence
5 pints of beer 10 whiskies
150 mg
Loss of self-control, exuberance, quarrelsomeness, slurred speech
6 pints of beer or 13 whiskies
200 mg
Stagger, double vision, memory loss
¾ bottle of spirits
400 mg
Oblivion, sleepiness, coma
1 bottle of spirits
500 mg
Death possible
600 mg
Death certain

Fig. 3.2

If you are caught driving with a blood alcohol level above 80 mg per 100 ml of blood you will automatically have your driving licence taken away for at least a year. *Long before* drinking enough to reach this level, reactions will be slowed down and accidents while driving become much more likely.

Fig. 3.3

Using Figs. 3.2 and 3.3, 2½ pints of beer or 5 glasses of whisky will take most people up to the legal limit. Half a pint of beer has the same effect as one glass of wine, so wine drinkers may have a maximum of five glasses if they are to be legally fit to drive.

● Problems

1. How many glasses of sherry will make someone drunk enough to be unable to speak clearly?

2. How many glasses of wine would have the same effect as 2 pints of beer and a double whisky?

3. What would be the blood alcohol level, after lunch, of a person who had one glass of sherry, two glasses of wine, and a brandy with his meal?

4. The blood alcohol level decreases naturally by about 15 mg an hour. If someone has a level of 200 mg, how long will it be before they are legally fit to drive?

5. At 2 a.m., after a birthday party, an 18-year-old has a blood alcohol level of 190 mg. Will he be fit to drive to work at 8.30 a.m.?

6. At 2 p.m., after a lunchtime outing from work, three friends have blood alcohol levels of 30 mg, 80 mg, and 130 mg, respectively. If they call in on the pub on the way home at 5 p.m., how much can each one drink and still legally drive home:

   (a) if they are drinking beer?
   (b) if they are drinking wine?

7. You can get approximately 15 glasses of sherry from one bottle. Draw a diagram like the one given in Fig 3.2 for whisky showing the effect of drinking sherry.

8. Repeat question (7) for table wine. (One bottle will give about 6 glasses of wine.)

9. Draw a graph showing the blood alcohol levels in the next 12 hours of someone who leaves a party with a level of 200 mg.

10. What other activities, besides driving, are dangerous after consuming enough alcohol to affect concentration?

4 Money

One of the basic problems in life is coping with and managing money. For example, money buys food, clothes, electricity, and many other things we want. That money may come from many different sources, e.g. wages, unemployment benefits, grants, etc. In this chapter we are concerned with a variety of such problems.

4.1. Premium bonds

Premium bonds provide a means of saving your money and at the same time entering a weekly and monthly draw for prizes ranging from £50 to £250 000. Bonds are in units of £1 but are sold in multiples of £5. The minimim purchase is £5, whilst for the wealthy there is a maximum holding of 10 000 units.

The list of prizes in Fig. 4.1 is taken from the back of a premium bond.

Example. How much money is given out in prizes of £500 or less each month?

750 at £500	=	£375 000
25 000 at £100	=	£2 500 000
75 000 at £50	=	£3 750 000
Total	=	£6 625 000

Prizes

The monthly prize fund is formed by calculating one month's interest, at the rate of 7 per cent per annum tax free, on each bond eligible for the draw. The fund is apportioned as follows (as from 1 April 1981).

Each week

1 prize of £100,000
1 prize of £50,000
1 prize of £25,000

Each month

1 prize of £250,000
5 prizes of £10,000
50 prizes of £5,000
250 prizes of £1,000
750 prizes of £500
25,00 prizes of £100
75,000 prizes of £50.

Provisions for the allocation of any residue or shortfall are contained in the prospectus.

Income tax
All prizes are free of United Kingdom income tax and capital gains tax.

Fig. 4.1

● Problems

1. What is the total prize money given out each year?

2. How much money is invested in premium bonds to enable prize money to be given? (Hint: use your answer to question 1 and the fact that the prize fund is formed by calculating the annual interest at a rate of 7 per cent per annum on each bond eligible for the draw.)

3. How many prizes are offered each year?

4.2. Time cards

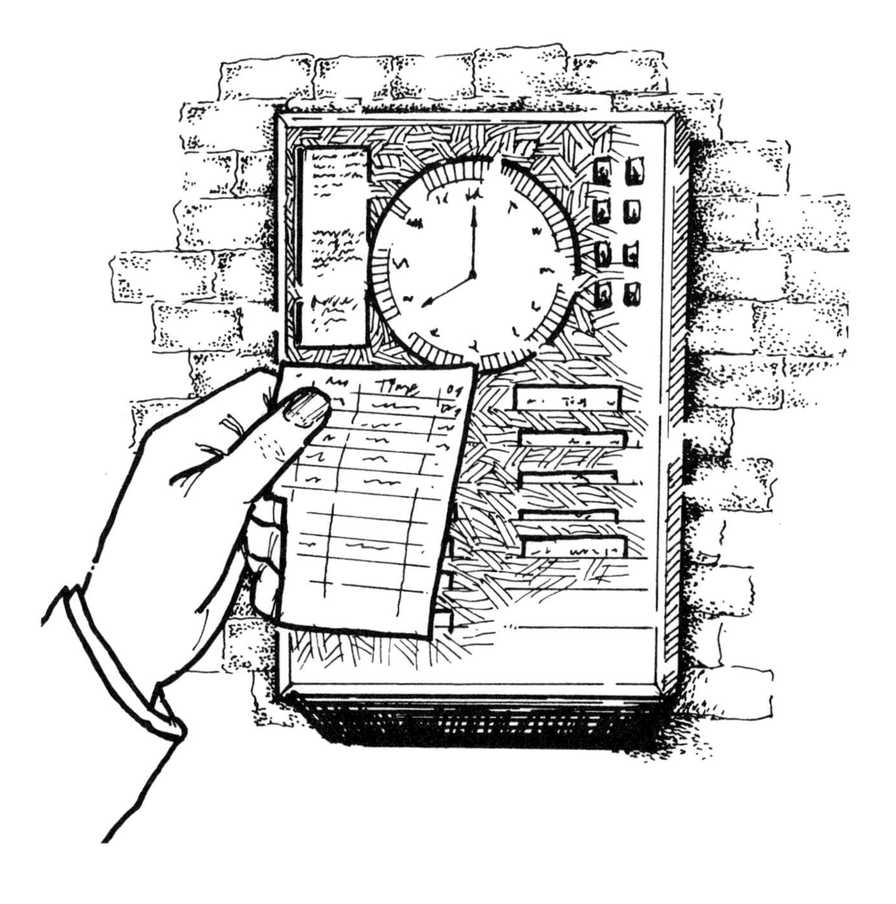

In many jobs you are paid an hourly rate, and so it is vital that management knows exactly how many hours you have worked in a week. A record of the time that you have been at work is kept using 'time cards'. Your time card is stamped when you clock in and when you clock off.

If you work more than a set number of hours in a week, you will be paid 'overtime' for the rest of your hours worked. This usually means that you are paid at a higher rate, for example 'time and a half'. If your usual hourly rate is £2.00, then 'time and a half' means £2.00 + ½ (£2.00) = £3.00.

In the following problem we look at a complete week's work for Mr Joe Smith.

● Problems

The time card for Joe Smith is shown in Table 4.1. His basic rate of pay is £2.00 per hour for a 35-hour week, overtime (except Sundays) is paid at 'time and a half', and any work on Sunday is paid at 'double time'.

Table 4.1

Name	Time	
Joe Smith	In	Out
Monday	08.30	17.05
Tuesday	08.21	17.15
Wednesday	09.15	18.30
Thursday	08.25	17.42
Friday	07.10	16.09
Saturday		
Sunday	09.12	12.42

Complete Joe's pay slip for the week. Gross pay means his pay *before* deductions for tax and national insurance are made.

Name				
Joe Smith	Basic	Overtime	Sunday	
Hours/mins				
Rate	2.00			Gross pay
Pay £				

In some problems it will not be quite so easy to calculate the pay. For example if you are paid at an hourly rate of £2.00, how much would you get paid for 5 hours 21 minutes work? For the 5 hours, you get 5 × £2.00 = £10.00, but it is not so easy to calculate the pay for 21 minutes. Since there are 60 minutes in an hour, we can write 21 minutes as the fraction 21/60 hours, and so the pay for 21 minutes is

21/60 × £2.00 = £0.70 or 70p (you might need to use a calculator here)

and the total pay is £10.70.

● Related problems

A. Here is Joe Smith's time card for the next week. Complete the pay slip.

Name	Time	
Joe Smith	In	Out
Monday	08.22	17.10
Tuesday	08.41	17.04
Wednesday	08.30	17.41
Thursday	08.52	16.52
Friday	09.10	17.51
Saturday		
Sunday	09.21	12.05

Name				
Joe Smith	Basic	Overtime	Sunday	
Hours/mins				
Rate £	2.00			Gross pay
Pay £				

B. If Joe had been paid at the standard rate of £2.20 per hour instead of £2.00 what would the gross pay have been for each week?

C. Complete the following pay slip (you will need a calculator).

	Basic	Overtime	
Hours/mins	36.00	4.23	
Rate £	1.72	2.12	Gross pay
Pay £			

4.3. Savings

Notices like the one above are a common sight nowadays – with government legislation ensuring that advertisements don't mislead the public – but they don't explain things very well. What interest *can* you expect on your savings?

Let's say that you're saving up to buy something really big – a stereo or a trail bike, maybe – and that you decide to put some money each week into a deposit account. Over a year or so this will earn quite a lot of interest and make the job of saving up a lot easier. What sort of deposit account should you choose?

The advertisement above actually refers to a deposit account with a building society, so let's start with that. To keep the numbers simple for the moment, let's assume you put £100 into the account and just left it there for a year. The phrase '12.14 per cent gross equivalent with income tax at 30 per cent' means that, before you pay any income tax on them, your savings will have earned 12.14 per cent in a year. So your £100 will have earned £12.14 interest, and your total savings will be £112.14.

Here comes the problem with building society accounts, though: even if you haven't got a job (and so don't pay income tax normally), the building society still charges you the 30 per cent tax. So your £12.14 interest is taxed at the 30 per cent rate, becoming

$$£12.14 \times 70/100 = £8.50 \text{ after tax.}$$

Thus, you started the year with £100 and now have £108.50 – an annual rate of 8.50 per cent, as in the advert.

There are plenty of other ways of depositing your savings – some of them are looked at below – but it's a good idea to check your local high street before deciding where to go, since the figures quoted change quite regularly.

● Problems

1. With the sort of deposit account mentioned above, you can get out as much of your savings as you wish whenever the building society is open. If you don't mind having to give a month's notice of a withdrawal, though, you can often get an even better interest rate.

(i) What would have happened, after a year, to the £100 if you'd put it into a higher rate account offering

13.93 per cent gross equivalent?

(ii) Or at 14.29 per cent gross equivalent?
(iii) What about the other way round? If the building society actually pays 8.75 per cent, how much would you have got if they hadn't charged you income tax?
(iv) Or at an annual rate of 10.75 per cent?

2. When you leave money in a deposit account for more than a year, the interest itself starts earning interest. With your original £100 at 8.50 per cent, you'd have £108.50 at the end of year one. If you left this alone for another year, you'd have compound interest with

$$£108.50 \times \frac{108.50}{100} = £117.72$$

at the end of year two – even more than the simple interest figure of £100 + £8.50 + £8.50. Of course, a simple way of writing this for one year is just

$$£100 \times 1.0850,$$

and for two years is

$$£100 \times 1.0850 \times 1.0850 = £100 \times (1.0850)^2.$$

For three years it would be

$$£100 \times (1.0850)^3,$$

and so on for any number of years.

(i) How much has a £100 principal become, at 8.50 per cent, after 5 years?
(ii) What about after 10 years?
(iii) How much more would you have after 5 years if you'd been able to get 8.75 per cent interest?
(iv) Or even 9.75 per cent?

3. On some types of account the situation is even more confused, because the building societies add in the interest *twice* each year. Take our original case of £100 at an annual rate of 8.50 per cent. The 8.50 per cent each year is clearly 4.25 per cent each half year, so the £100 becomes

$$£100 \times (1.0425)^2 = £108.68$$

if the interest is added in twice each year.

(i) How much would the £100 have grown to after 2 years on this scheme?
(ii) What about £300 for 5 years?
(iii) Now consider an annual rate of 8.75 per cent with this scheme. What about the £300 for 5 years? Or £500 for 10 years? Or £1000 for 15 years?

● Related problems

A. If you want to be able to withdraw money as easily as possible, a National Savings Bank ordinary account is probably best – it can be used at any post office in the country, and these are open six days a week. The disadvantage is that it only pays 5 per cent interest!

The rules state that the first £70 of interest is free of income tax completely. How much can be held in a NSB ordinary account before you would start paying tax.

B. The Big Four high street banks are very familiar – Lloyds, Barclays, National Westminster, and Midland – but there are also many other smaller ones. They all offer very similar deposit facilities – a current account (with cheque book, but no interest) or a deposit account (with interest, but less access to your money).

The current interest rate in one high street bank is 9 per cent for accounts up to £1000, but 10 per cent for accounts over that figure. You have to pay tax on this interest, though, if you are a tax-payer. What are the true rates when income tax has been taken into account?

C. Access cards are a very convenient way to pay bills nowadays – no money changes hands, but you have to show your Access card and sign the slip instead. A bill is presented at the end of every month, to cover all the items purchased using the Access card, and interest is charged at the rate of 2¼ per cent per month. What is this as a yearly interest rate?

4.4. Cash help

When you leave full-time education your parents can no longer claim child benefit or get tax relief for you as a dependent relative. You are expected to support yourself. If you start work, your employer will pay you, but, if you cannot find a full-time job, you can claim supplementary benefit at your local social security office.

You claim on the leaflet *Cash help*, part of which is shown here. (1983 figures).

What is supplementary benefit?

It is cash you can get if the money you have coming in is less than you need to live on.

You can't get it if you are in full-time work. But if you have children you may be able to get FIS (Family Income Supplement) instead. Get leaflet FIS.1 from a post office or social security office.

You can claim supplementary benefit if you are:

- over pension age
  or
- unfit for work
  or
- bringing up children on your own
  or
- unemployed and can't get work
  or
- working only part-time
  or
- needed at home to look after a disabled relative.

Did you know:

- You don't need to have paid national insurance contributions.
- Savings up to £2000 won't affect your benefit at all.
- Owning your own home won't stop you getting benefit.
- If you get supplementary benefit, you and your family can get other things free – like school meals, milk for the under-fives, visits to the dentist, glasses and prescriptions.

If you want to know more:
Ask at your social security office. Or go to a Citizens Advice Bureau or other advice centre. Detailed information is given in *Supplementary Benefits Handbook*. You can see this at your local library.

How is it worked out?

- First we work out which sum of money in the table below applies to you.

	If you are under pension age	**If you are over pension age**
Married couple	£37.75	£47.35
Single person paying rent or owner-occupier	£23.25	£29.60
Other person		
aged 18 or over	£18.60	£23.65
aged 16–17	£14.30	

- Then we add on for each of your children
  - aged 11–15 £11.90
  - aged under 11 £7.90
- Then we add on for
  - blind people £1.25
  - people aged 80 or over £0.25
- Then we add more for your housing costs:
  - rent and rates for your home

    or
  - your rates, mortgage interest and an allowance to help with insurance and maintenance if you own your home
- but if some people in your home aren't dependent on you, we add less or none at all.
- Or if you live in someone else's household we add £2.55.
- More is added for extra expenses such as a special diet, extra heating ot central heating.
- When we have added together all these needs we count the money you already have coming in. If it is less than your needs, we will pay you supplementary benefit to make up the difference.
- But if you are signing on, and are unemployed through your own fault, you may get less benefit.
- Even if your income seems to be a little more than your needs, you should still claim: we may not have to add up all your income, or you may have some extra needs, so you may still be able to get some benefit.

From the information given, a 16-year-old living at home would get £14.30 + £2.55 = £16.85.

● Problems

1. How much should a 19-year-old living at home expect to get?

2. How much is allowed for a married couple over pension age?

3. An elderly man aged 85 lives with his daughter and her family. How much supplementary benefit will he get?

4. A single parent has two children aged 12 and 14. The rent and rates of their home are paid as part of their supplementary benefit. How much cash is to be expected on top of this?

5. A married couple with no other income has savings of £1800 in the building society. Will this affect their benefit?

6. A widow lives with her nine-year-old daughter and 13-year-old son in a house which is completely paid for. How much benefit will she get?

7. A young married couple on supplementary benefit have a baby. How much extra money will they get each week?

8. An 18-year-old boy living at home is paid £8.50 a week for a Saturday job. If he has no other income how much supplementary benefit will he get?

9. A married couple with children aged 10, 12, and 14 received supplementary benefit for a total of 8 weeks. How much money did they get altogether?

10. Three 17-year-olds live in the same town. They all live at home with their families and each is getting supplementary benefit. If they moved out and shared a flat together, how would their benefit be affected?

5 Home

Everyone needs a place to call 'home' and in the home there are many occasions when simple mathematics is used to solve household problems. For example, an estimate of the electricity bills requires the ability to read the 'numbers' on a meter and to multiply the quantity of units used by the price of a unit; this may involve the interpretation of an electricity board's pricing code.

It is in their middle- to late-teens that many people begin to think about moving away from their parents and setting up their own home. The choice of what type of accommodation may be wide, e.g. it might be digs or it might be house purchase. Having found a place of their own, a young person may have to furnish it or decorate it and will certainly have to pay for running it, i.e. heating, lighting, etc.

This chapter gives the student an insight into some of the decisions and planning needed when setting up their own home. Some of the problems in § 5.1 may be suitable for a class discussion.

5.1. Independence is expensive

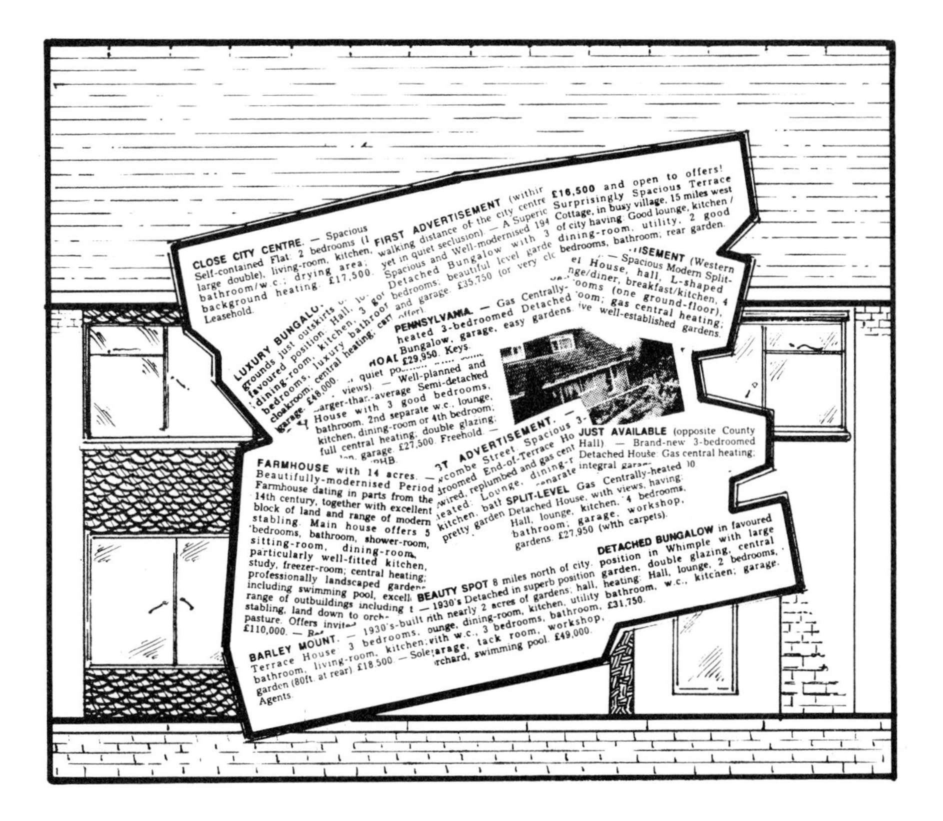

Fed up with living at home? Time to find a flat of your own? There comes a time in most teenager's lives when there is a desire to leave home. You may want to leave home because you don't get along with your parents or because you don't get the freedom or independence you want. However, setting up your own home can be expensive and many of the familiar 'home comforts' are gone. Looking at the advertisement above, is that all it really costs?

In the problems we consider the various types of accommodation available and some of the decisions that have to be made.

● Problems

1. What are the different types of accommodation available and what facilities do you get with each type? What does each type of accommodation cost in your area?

2. The rent is only part of the story. If you move into an unfurnished flat or house then you will need to buy furniture. Go and find out about the cost of setting up home.

3. Now having paid the rent and furnished your house or flat, it costs money to heat it, to cook your meals, and so on. What are these 'running costs'?

● Related problems

A. Suppose that you have built the house of your choice described in Chapter 6. Investigate the cost of furnishing it.

5.2. Furniture moving the easy way

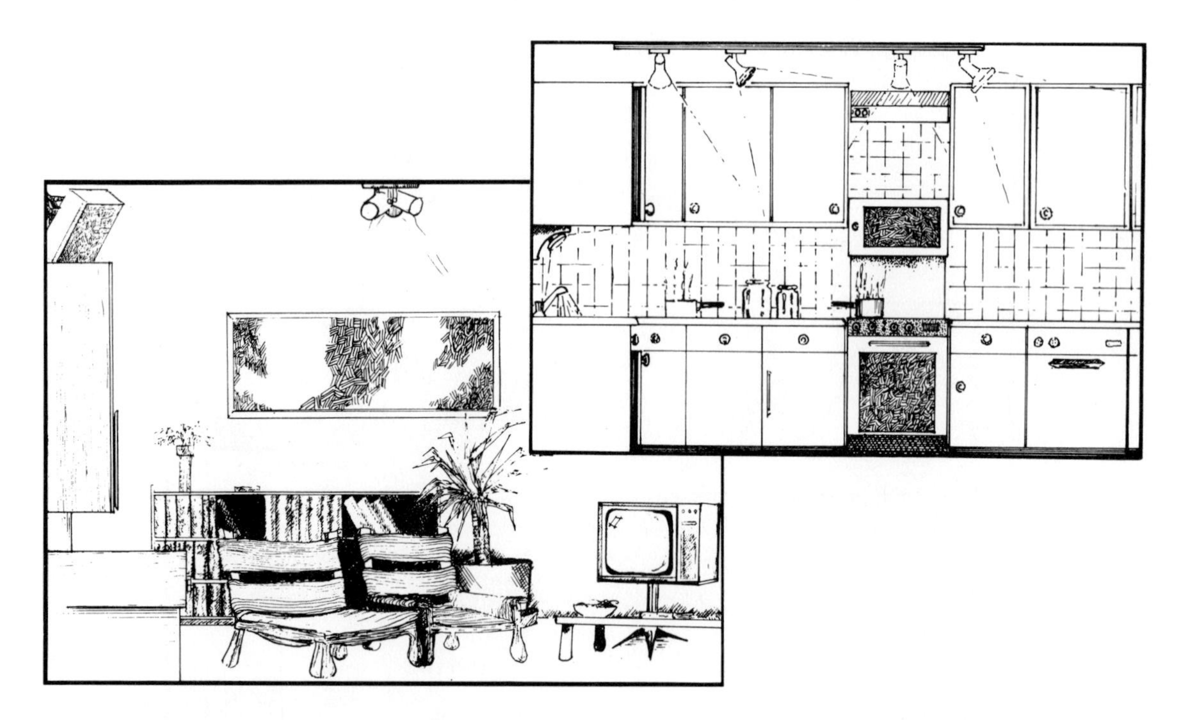

You have a small room and a selection of furniture to put in it. You cannot get all the furniture into the room, so you have to choose which pieces you want and decide if they will fit into the room. The room is so small that it is difficult to keep moving pieces of furniture around in it. However, if you make a scale plan of the room and scale drawings of the furniture you can move pieces of paper, instead of furniture, around.

If you feel that you need some practice in scale drawings, try the examples and exercise at the end of this section.

● Problems

1. Imagine that you are moving into a new house. Your bedroom is shown in Fig. 5.1. Take a sheet of graph paper. Using a scale of 1 mm to 1 cm, draw a plan of this bedroom, marking on it the doorway and the window. (You will then have a plan that is one-tenth real size.)

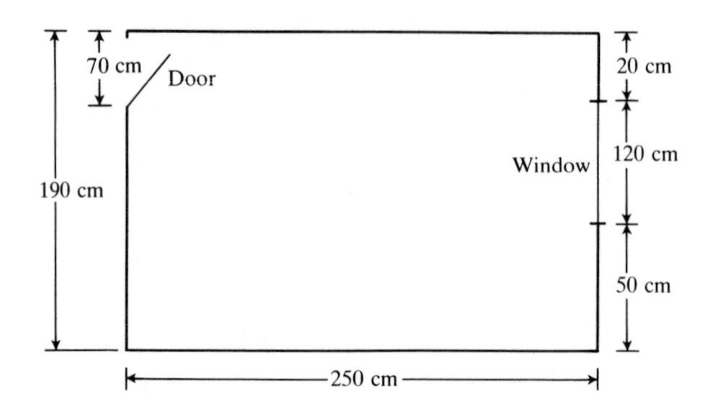

Fig. 5.1

2. Your mother says that there are several pieces of furniture that you can have if you want. (Table 5.1). They will not all fit into the bedroom so you must choose what you want and decide where you want it to go. It would be hard work moving furniture in and out of the bedroom while you tried to make up your mind, so you are going to use the plan you have drawn to help you decide.

Table 5.1

Furniture	Size	Furniture	Size
Bed	90 cm by 185 cm	Dressing table	75 cm by 35 cm
Wardrobe	90 cm by 50 cm	Small armchair	46 cm by 46 cm
Small cupboard	48 cm by 51 cm	Circular stool	30 cm in diameter

Take a fresh sheet of paper. Using the same scale of 1 mm to 1 cm, draw plans of each of the pieces of furniture in Table 5.1 and cut them out. Write on them what each piece represents (e.g. bed, dressing table, etc.). By moving them around the plan of the bedroom, decide which pieces of furniture you want and where you will put them.

3. Look at the plan you have made with the furniture in place and answer the following questions.

(a) If you have chosen to use the wardrobe can you open its doors in this position?

(b) If you have chosen to use the dressing table can you open the drawers in this position?

(c) If you want to sit down in front of the dressing table, can you do so?

(d) Is there room for the bedroom door to open?

(e) In the winter the cold air drops straight down from the window. If you have the bed under the window, do you mind?

(f) Is the furniture you have put in front of the window going to block out the light?

(g) Now you have considered these points do you want to try another arrangement of the furniture? If you do, then do so.

4. Mark in the final position of the furniture on the plan of your bedroom.

● Related problems

A. Fig. 5.2 shows the positions of the 2 electric sockets (□) and the only light (*) in the room shown in Fig. 5.1.

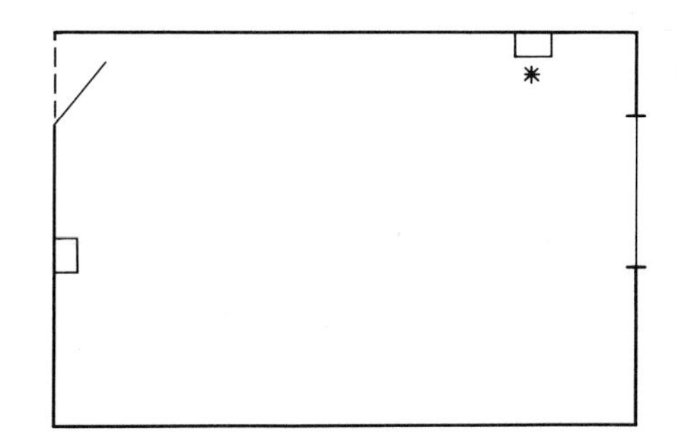

Fig. 5.2

If you knew this would it alter your arrangements? Would you want to add other items, not on the original list? If so, measure the items you want to use, make scale drawings as before, and replan the room.

B. Repeat the original exercise with a bedroom that measures 270 cm by 300 cm. With this larger bedroom is there any other piece (or pieces) of furniture that you would like to have? If there is, measure its size in cm and make a scale drawing of it and use it as well as the original pieces of furniture.

C. Make a plan of your own bedroom at home and use the same method to rearrange your own furniture.

● Examples on scale drawings

1. Using a scale of 1mm to 1 cm draw a line to represent a line of 55 cm long.

If 1 mm represents 1 cm, then 55 mm represents 55 cm.

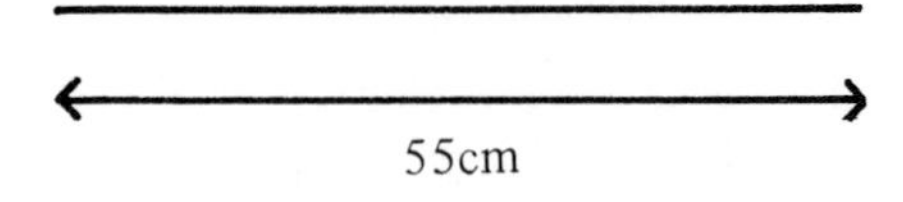

2. Using a scale of 1 cm to 5 cm, make a scale drawing to represent a rectangle 15 cm by 40 cm.

If 1 cm represents 5 cm,
then 3 cm represents 15 cm (15 cm = *3* $\times$ 5 cm),
and 8 cm represents 40 cm (40 cm = *8* $\times$ 5 cm).

● Exercises

1. Use a scale of 1 mm to 1 cm to represent the following lengths

(i) 50 cm; (ii) 100 cm; (iii) 45 cm; (iv) 38 cm; (v) 72 cm.

2. Use a scale of 1 cm to 4 cm to represent the following lengths

(i) 8 cm; (ii) 12 cm; (iii) 20 cm; (iv) 36 cm; (v) 10 cm.

3. Using a scale of 1 cm to 5 cm, make scale drawings of squares with the following sides

(i) 10 cm; (ii) 20 cm; (iii) 30 cm; (iv) 25 cm; (v) 12½ cm.

Now get this exercise checked by your teacher before going on to the problems.

5.3. Ready pasted?

You are going to do some wallpapering. In order to calculate how many rolls of wallpaper you need you may use Table 5.2. Wallpaper shops supply tables like this one free of charge. First you have to measure the height of the room from the skirting up to where the wallpaper will finish. Then find the distance round the edges of the room, including doors and windows. Now you read off from Table 5.2 the number of rolls required. All measurements are given in metres.

Table 5.2

Height from	Number of rolls required for measured wall length* (m)												
skirting (m)	8	9	10	11	12	13	14	15	16	17	18	19	20
2.0	4	4	4	5	5	6	6	6	7	7	8	8	8
2.2	4	4	5	5	6	6	7	7	7	8	8	9	9
2.4	4	5	5	6	6	7	7	8	8	9	9	10	10
2.6	5	5	6	6	7	7	8	8	9	9	10	10	11
2.8	5	6	6	7	7	8	8	9	9	10	10	11	12
3.0	5	6	6	7	8	8	9	9	10	11	11	12	12

* Wall lengths include doors and windows.

EXAMPLE ON READING FROM TABLE 5.2

Calculate the number of rolls required if the height above the skirting is 2.2 m and the distance round the room is 12 m. Look down the left-hand column of Table 5.2 until you reach 2.2 m. Move along this row until you are under the 12 in the top row (see Table 5.3). This tells you that you need six rolls.

If you need more practice in reading from Table 5.2 try the exercises given at the end of this section.

Table 5.3

Height from skirting (m)	Number of rolls required for measured wall lengths* (m) 8	9	10	11	12	13	14	15	16	17	18	19	20
2.0	4	4	4	5	5	6	6	6	7	7	8	8	8
2.2	4	4	5	5	6	6	7	7	7	8	8	9	9
2.4	4	5	5	6	6	7	7	8	8	9	9	10	10

* Wall lengths include doors and windows.

● Problems

1. Your bedroom is rectangular and measures 4 m by 3 m. The height above the skirting is 2.2 m and you have chosen wallpaper costing £3.50 per roll. Find the distance round the walls of your bedroom and read off from Table 5.2. How many rolls will you need? How much will this cost?

2. The spare room is also rectangular and measures 3.4 m by 3 m. Calculate the distance round the room. How do you deal with the problem that this number is not in Table 5.2? Now work out how many rolls of wallpaper you need, assuming a height of 2.2 m above the skirting as before.

How much will the wallpaper cost if the price of a roll is £2.85?

3. Mum and Dad decide to wallpaper their room with woodchip, at £1.15 per roll. If their room measures 4.2 m by 3.5 m, with the same height above skirting as before, find the number of rolls of wallpaper required. How much will they have to pay for them?

If they paint the woodchip, 2 litres of paint are required. How much will this cost, if 1 litre costs £2.70? What is the total cost of decorating the bedroom?

4. Having decorated upstairs you now want to decorate the lounge, which is L-shaped, and is shown in Figure 5.3.

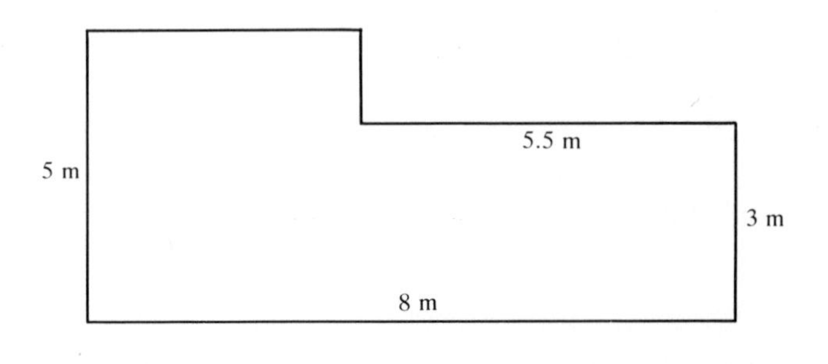

Fig. 5.3

(a) What is the distance round the room? Assuming the height of this room to be 2.4 m, can you read the number of rolls required from Table 5.2? If you cannot, find how many rolls you would need to decorate half the room, then double your answer. Will you have enough? If you choose paper at £5.30 per roll, how much will it cost?

(b) Instead, it is decided to use this paper for the longest wall of the lounge, but to use the same woodchip and paint as in Mum and Dad's bedroom for the other walls. How much will it cost to decorate the lounge now? How much cheaper is the decorating this way?

● Related problems

A. The table below is taken from the back of a wallpaper paste packet:

APPLICATION	Quantity of cold water to use	Approximate coverage*
Normal wallpapers including woodchip	16 pints	11–13 rolls
Washable and Vinyl wallcoverings	13 pints	10–11 rolls
Novamura	10½ pints	8 rolls
Embossed papers (e.g. Anaglypta)	10½ pints	4–5 rolls
Heavy embossed papers (e.g. Supaglypta)	9 pints	3–5 rolls
Polystyrene tiles (not veneer)	9 pints	180 sq. ft.

*NB: These figures are intended as a guide only.

How many packets of paste will you require for the redecoration?

B. Table 5.2 does not take account of doors and windows. In the spare room the door is 2 m tall and 83 cm wide. The window is 135 cm wide and 80 cm high.

Area of door = 200 cm × cm = cm²
Area of window = cm × cm = ______ cm²
Total area of paper wasted = cm²

If the wallpaper is approximately 50 cm wide, then

the length of wallpaper wasted $= \frac{\quad}{50}$ cm
= cm
= m

If a roll of wallpaper is approximately 10 m long, roughly what fraction of a roll are you wasting in this case?

C. Measure the distance round your own bedroom at home. Measure its height from the skirting and calculate the number of rolls of wallpaper required to redecorate it.

D. How would you allow for using a patterned wallpaper?

● Exercises

1. Complete the following table which gives height from skirting and measurements of walls (rounded up where necessary). We've completed the first line for you.

Height from skirting (m)	Measurements of walls (m)	Number of rolls
2.2	12	6
2.6	18	
2.0	9	
3.0	14	
2.8	19	
2.4	10	
2.6	11	
3.0	18	

2. Assuming a height above skirting of 2.4 m, what is the longest length of walls you could paper with: (a) 6 rolls; (b) 7 rolls; (c) 8 rolls?

3. If the distance round the outside of the room is 15 m, what is the largest height from skirting, so that you could paper the room, with: (a) 7 rolls; (b) 8 rolls?

5.4. Tape it

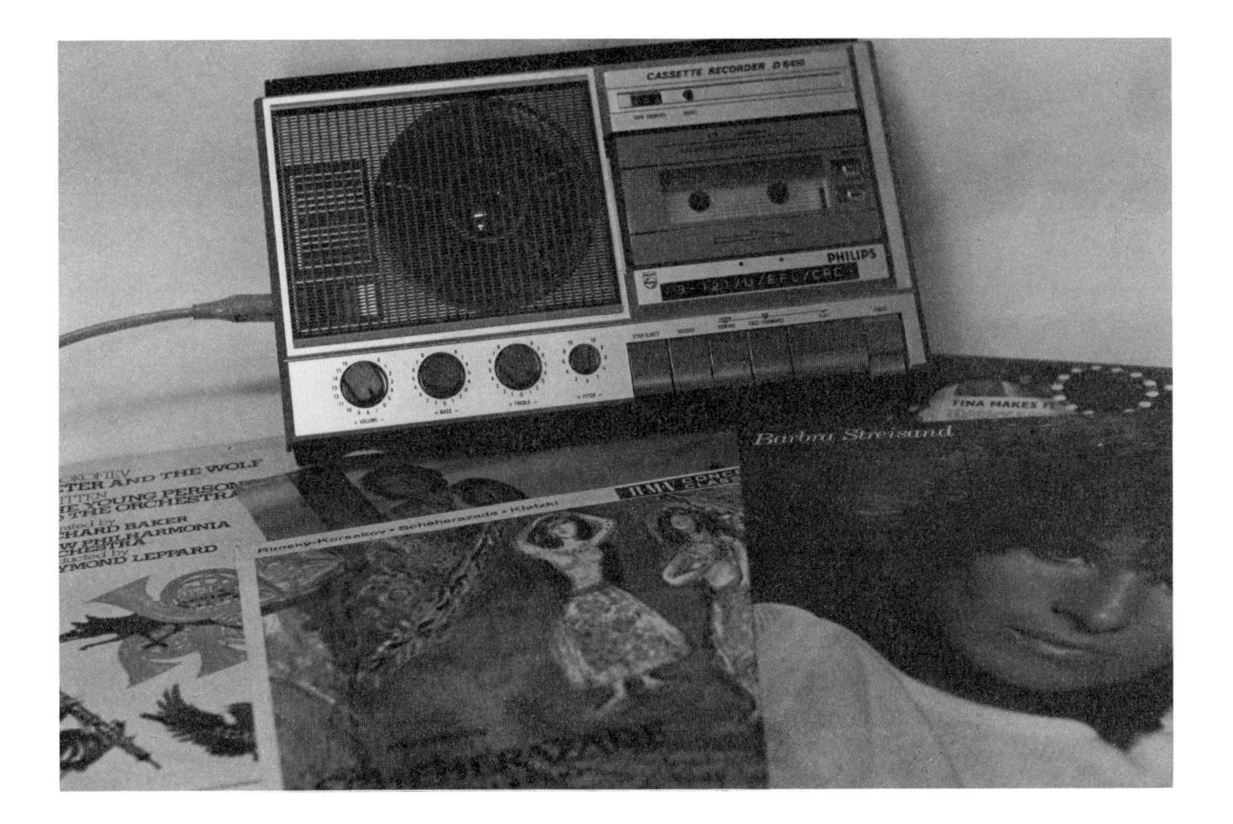

Suppose that you are making a tape for background music at a party. You've got an LP from which you want to take 20 minutes of Side 1 and 18 minutes of Side 2, and some singles lasting 2, 2½, 3, 3½, 4, 4½, and 5 minutes.

You want to make up a 60-minute tape with as little unused tape as possible, and you don't really mind if not all the singles are used.

One way of doing this is to put

$$18 + 2\tfrac{1}{2} + 3\tfrac{1}{2} + 5$$

on one side of a C60 cassette, and

$$20 + 2 + 3 + 4$$

on the other side. This doesn't seem a very good solution though, since neither side of the tape is full.

● Problems

1. Is it going to be possible to record everything on a C60 cassette?

2. Can you find a combination which leaves *no* blank tape?

3. If you used another LP as well, from which you wanted 19½ minutes of one side and 21 minutes of the other, could you make up a 90-minute C90 tape?

4. C60 tapes (with 30 minutes on each side) cost £1.15, while C90 tapes (with 45 minutes each side) cost £1.55. What is the cheapest way of recording 180 minutes of music?

5. You want to tape all the records shown in Fig. 5.4 as cheaply as possible. How many tapes will you need? How will you arrange the music on the tapes?

Warning: Remember that the track times given are in minutes and seconds, *not* minutes and decimal parts of minutes. Even though both the Streisand and Dylan records give times like 3.42, this is 3 minutes and 42 seconds, *not* 3.42 minutes.

Fig. 5.4

6 House design

The design and building of houses is an area which affects everyone, particularly teenagers who are trying to accommodate their own life-style within the confines of the parental home. It is also one which many people take for granted unless and until they become involved in it, and one which is rich in mathematics.

6.1. House design

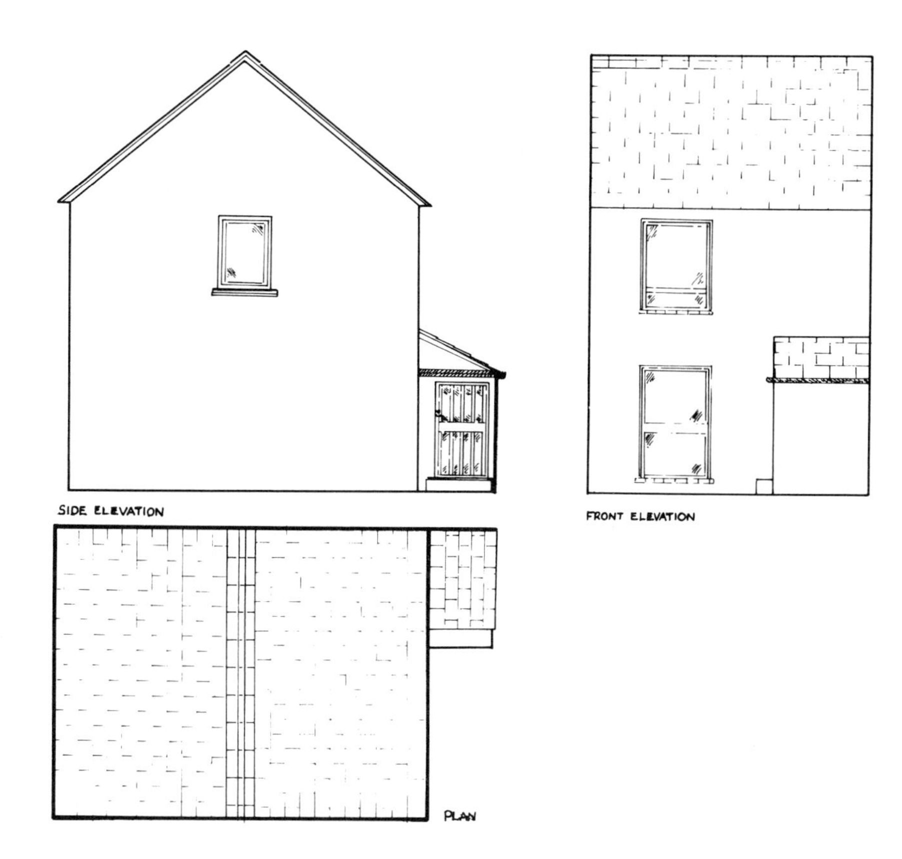

THE PROBLEM

You have the opportunity to design and build a house. How do you set about it? As a start here are some questions which need to be answered.

1. What sort of house? Detached, semi, bungalow, town house? What accommodation should it have? What rooms and how big should they be?
2. What is the site like on which you are going to build?
3. What structural problems do you need to consider when designing the layout of the rooms?
4. What technical problems (heating, lighting, ventilation, plumbing) do we need to consider?
5. What style is the outside of the house going to be? Are there any planning regulation restrictions on this?
6. How much is it going to cost to build? How do we make it cheaper when it is going to cost more than we can afford?
7. How long will it take to build.

As this is a very large problem, it is broken down into stages over the next few pages.

6.2. Describing houses

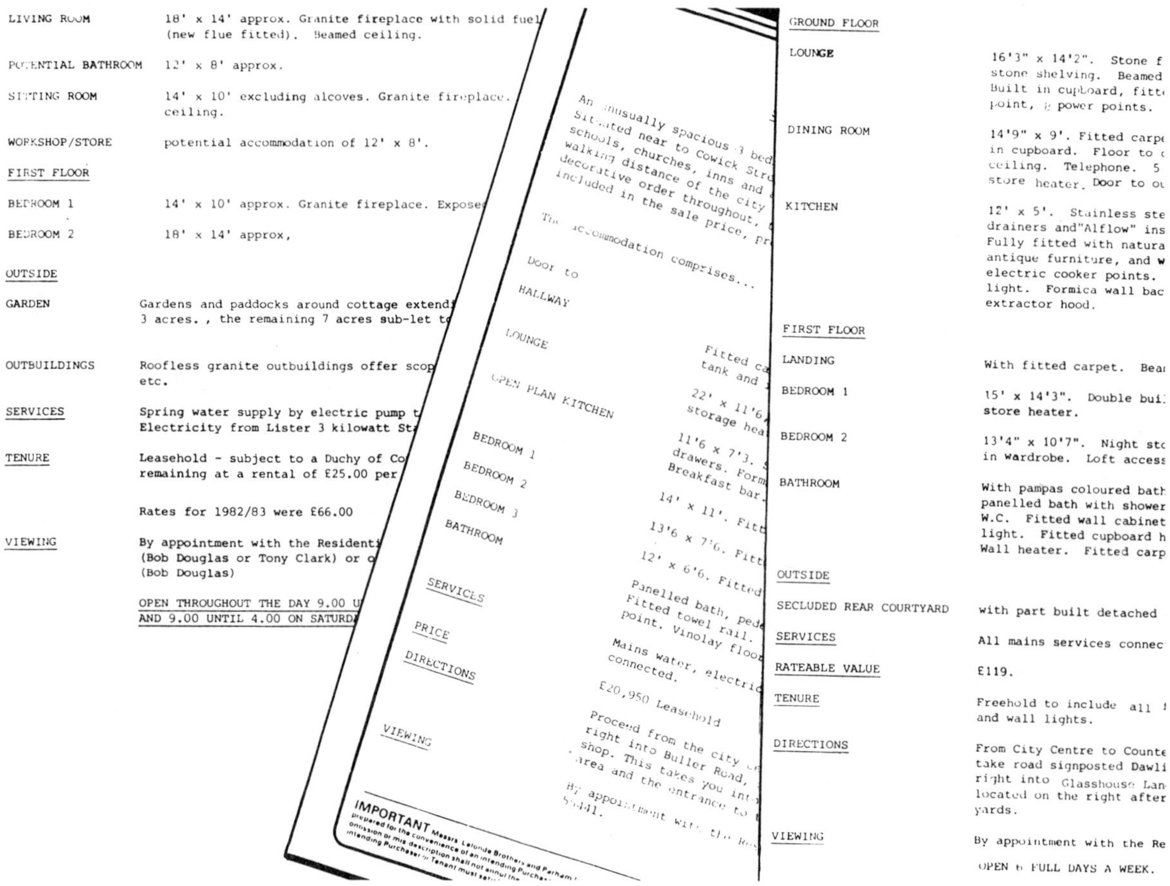

Fig. 6.1

When you are selling a house, it has to be described so that possible buyers have some idea of what the house is like. The council housing department also have to write down various things about each of their houses, so that when houses become vacant they know for which of the people on their waiting list it would be suitable. Figure 6.1 shows the sort of advertisement which estate agents hand out.

● Problems

1. Describe you own house (or a friend's or a relative's) as it might be described by an estate agent.
2. How would the council housing department describe the same house?

● Related problems

All the problems in this chapter link together to produce the design of a house.

A. What sort of house should it be?

You need to answer some simpler questions.

(i) Is the house going to be attached to others? Unless you are building your house as part of a co-operative effort it will probably be detached.

(ii) How many floors are there going to be:
One-storey bungalow;
Two-storey house;
or three-storey town house?
The more floors there are the more stairs you will have to climb, but the more garden you will have, as the building will occupy less of the plot of land.

(iii) Is there going to be a garage, and, if so, is it going to be detached from the house, attached to it, or built into it?

When you have thought about these questions, write a few sentences describing the type of house you are going to design.

B. What accommodation should it have?

You need to consider:

(i) the number of bedrooms;

(ii) the number of living rooms;

(iii) whether the WC should be separate from the bathroom;

(iv) whether you want any non-standard rooms, e.g. downstairs loo, utility room, conservatory, balcony.

Write a few sentences describing the accommodation that your house will have.

6.3. How well do you know your house?

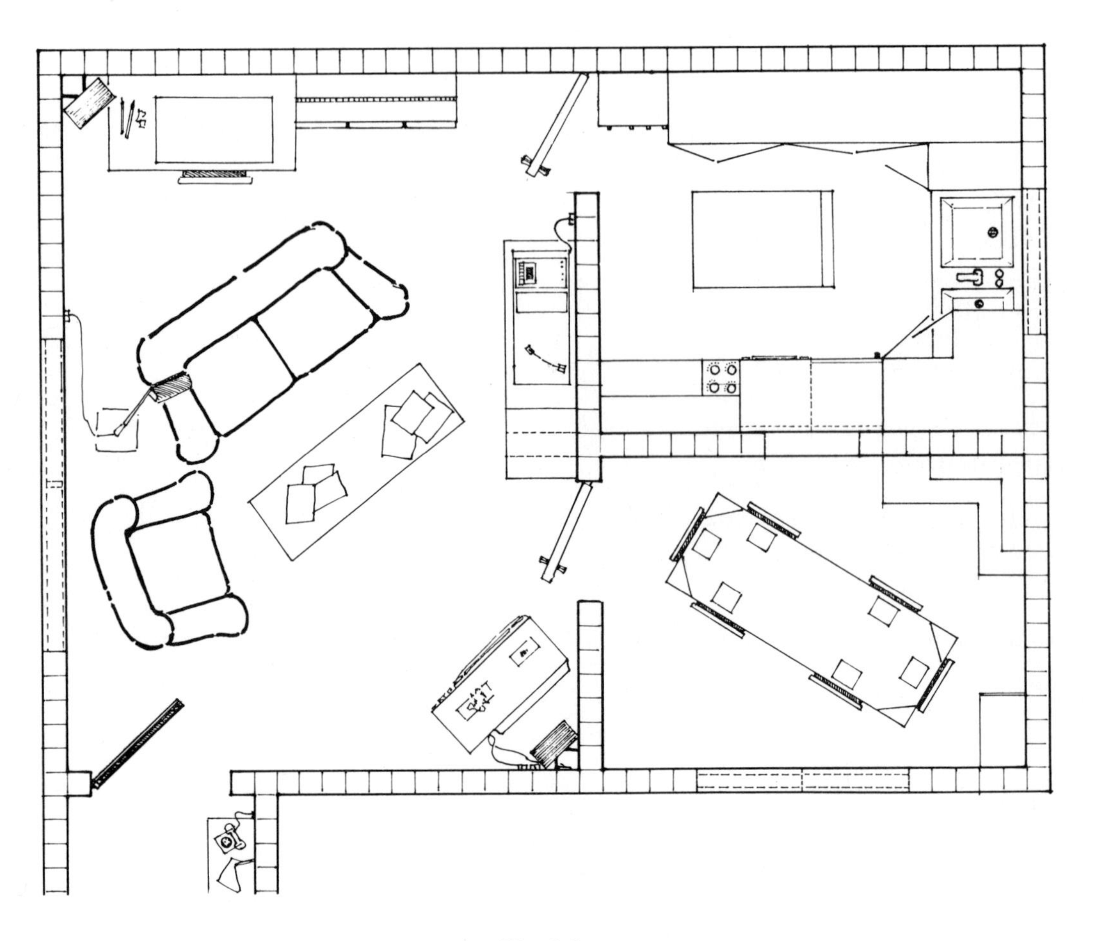

Fig. 6.2

It is a common human failing to take very familiar things for granted and not really notice them.

● Problems

Try to draw a plan, like the one in Fig. 6.2, of your house (or another that you know well). Use squared or graph paper and draw it so that each square represents approximately a metre. Do both downstairs and upstairs (unless you live in a bungalow or flat!).

How close were you?

Draw a more accurate version of the plan when you get home and can look at the house again.

6.4. Building costs

Table 6.1 gives approximate figures for the cost (in £/m²) of building various types of house. To get an estimate of the cost of building a house:

1. Calculate the total floor area (all floors);
2. Multiply this area by the appropriate figure from Table 6.1.

Thus,

$$\text{Total cost} = \text{cost/m}^2 \times \text{ground floor area.}$$

This gives the basic cost of building the house not including things like heating, lighting, and power points and any additional features not normally included in a 'basic house'.

Table 6.1

Region	Building costs (£/m²) for type of house			
	Semi	Detached	Terraced	Bungalow
London	390	420	380	430
South East and Scotland	355	380	345	390
Wales, North East and Yorkshire	335	355	325	370
Rest of the country	325	345	315	355

● Problem

Use Table 6.1 to calculate the cost of building your own house, or a friend's house.

● Related problems

A. Work out the cost of the house you have designed.

B. If you had to save £2000 on the building cost, how would you alter your design? You can describe your solution either in words or by a sketch plan.

6.5. Heating and lighting

Most houses built in this country nowadays have central heating and all have electric light and power points. The places where you have radiators, lights, and power points can make things very much more (or less!) convenient.

● Problems

1. Before deciding where to put radiators in each room, you have to decide how big they need to be. As a rough guide, a room needs 1 m length of standard radiator for every 5 m^2 of floor area. (You can get double radiators as well, in which case you only need half the length.) Choose the plan of your house, or your friend's, and make a table like this.

Room	Area m^2	Single/double	Length (m)
Dining room	16	S	3
Hall and stairs	8	D	1

2. Trying to show the position of radiators on a plan could make it confusing, so we might put them on an *overlay* instead.

Lay a piece of tracing paper over the plan and draw in the outline of the *outside* walls only. Now draw in the radiators to scale and show also where the central heating boiler is to go.

3. Draw another overlay showing the position of lights and power points.

● Related problems

A. Work out problems 1, 2, and 3 for your own design.

B. Costs:

central heating boiler and pump	£400
radiators – single (incl. pipework)	£30 per metre length
– double (incl. pipework)	£70 per metre length
fuse box + meter installation	£250
power points (double) (incl. wiring)	£25
ceiling lights (incl. wiring)	£15
wall lights (incl. wiring)	£20

Use these figures above to estimate the cost of the heating and lighting installation in the house you have designed.

6.6. What next?

Building a house, like many other large projects, consists of a large number of smaller jobs, done by different specialists, which have to be done in the correct order. For example, the walls cannot be built before the foundations are laid, the wiring must be put in before the plastering is done, and so on. In order to plan when various jobs will need to be done, we have to have a way of representing the order in which things must be done.

As an example, Table 6.2. gives the tasks involved in making a cup of tea. We can draw a network of lines to show these connections (Fig. 6.3).

Table 6.2

	Operation	Operations which must be done first
A	Boil kettle	–
B	Get cup	–
C	Put milk and/or sugar in cup	B
D	Get pot	–
E	Put tea (bag) in pot	D
F	Pour water in pot	A, E
G	Let tea brew	F
H	Pour tea	C, G

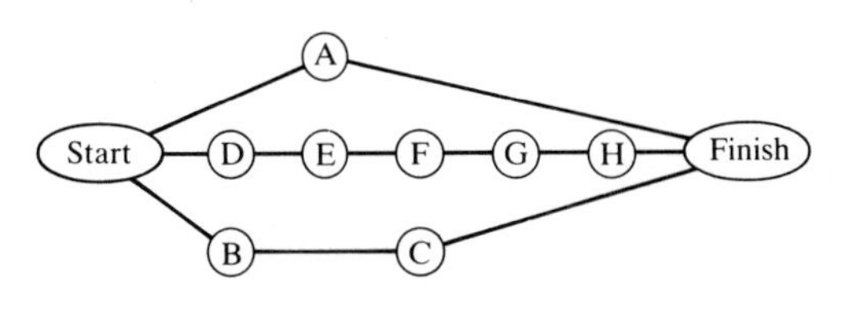

Fig. 6.3

● Problems

Make a table of operations for making beans on toast and draw a network like the one in Fig. 6.3.

The network diagrams are useful for finding out how long the whole job will take. For example, the operations in making a cup of tea might take:

Operation	A	B	C	D	E	F	G	H
Time taken (seconds)	120	10	30	10	10	10	90	5

How long, overall, does it take to make a cup of tea? If we add all the times together, this will be too long because other things can be done while the kettle is boiling.

If we put the times into the network, we get Fig. 6.4. Now the problem comes down to finding the route through the network from 'Start' to 'Finish' which gives the longest total time.

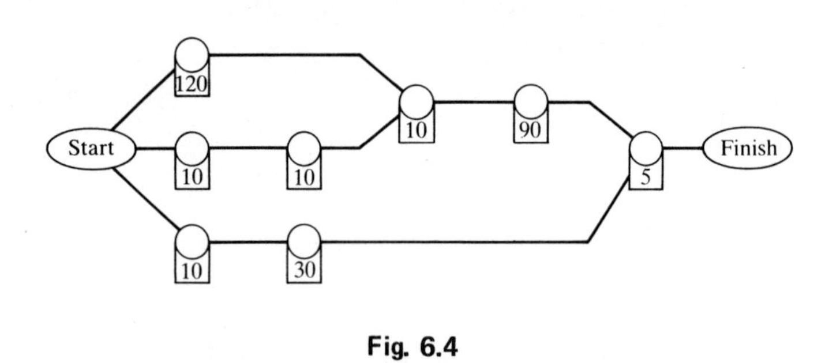

Fig. 6.4

Thus, take each possible route through the network, and in turn add up the times of the operations on the route. Find the largest total. This route through the network is called the *critical path*; any operations on this route must be started as soon as they can or the whole job will take longer.

The possible routes through the 'tea' network (Fig. 6.4) are

A F G H taking $120 + 10 + 90 + 5 = 225$ seconds
D E F G H taking $10 + 10 + 10 + 90 + 5 = 125$ seconds
B C H taking $10 + 30 + 5 = 45$ seconds.

Thus, the critical path is A F G H and the whole job will take 225 seconds, i.e. 3¾ minutes.

Add estimates of the times for the various operation in 'cooking beans on toast' to your network. Work out the critical path and the total time taken.

● Related problem

Work out how long you think it will take to build your house.

Procedure:

Write down a list of operations, giving each a letter; alongside each, write the letters of any other operations which must be completed before this one can be started; alongside each, write your estimate of how long you think the operation will take.

Then,

Draw the network;
Put in the times;
Find the critical path and the time that the whole job will take.

7 Buying

The idea of 'shopping around' is important when buying different articles. Shops can always be found that advertise *discounts*.

For example, the same 16 inch colour television set may vary in price by as much as £100 (from £225 to £325) between different shops. However, there is more to purchasing articles than finding the cheapest price. The after-sales service provided by different shops varies quite considerably. A local electrical shop may provide a better after-sales service than a large discount store. So what is *a good deal*?

Furthermore, to pay out £300 at one time is perhaps not always possible so that *hire purchase* is a real alternative.

This chapter deals with problems involving buying some articles which might be of interest to the students and the cost of advertising articles for sale.

There may be several points to discuss with the class before tackling the problems given. When buying by instalments, for example, the amount of *interest* is an important quantity to explain.

Some discussion points for this chapter might be

(a) What constitutes a good deal?

(b) Are cheap groceries better/worse quality than 'branded names'?

(c) What is deposit?

(d) What is interest?

(e) Talk about different periods of repayment and different instalment amounts.

(f) Are there any other sources of borrowing?

7.1. How much off?

Bill wishes to buy a new car marked at £3600 where the garage offers a discount of 12½ per cent off new cars with no part exchange. How much will he pay if he has no car to part exchange?

The problem can be tackled in several ways.

METHOD 1

Treating £3600 as 100 per cent it is possible to find the values of 10 per cent and 2½ per cent which can then be added to find the total discount.

10 per cent of £3600	=	£360
5 per cent of £3600	=	£180
2½ per cent of £3600	=	£90
Total discount	=	£360 + £90
	=	£450

Selling price = £(3600–£450) = £3150.

METHOD 2

The total discount can also be found by finding 12½ per cent of £3600 and subtracting it from the marked price.

Discount on £3600 = £3600 X 12½/100 = £450
Selling price = £(3600–£450) = £3150.

METHOD 3

The selling price can be found directly by treating the market price, £3600, as 100 per cent and evaluating the selling price as a percentage.

Selling price per cent = (100–12½) per cent
= 87½ per cent.
Selling price = £3600 X 87½/100
= £3150.

METHOD 4

The problem can also be solved graphically (Fig. 7.1). A discount of 12½ per cent means a saving of £12.50 on every £100 or £125 on every £1000.

Discount on £3600 is £450
Selling price = £(3600–450) = £3150.

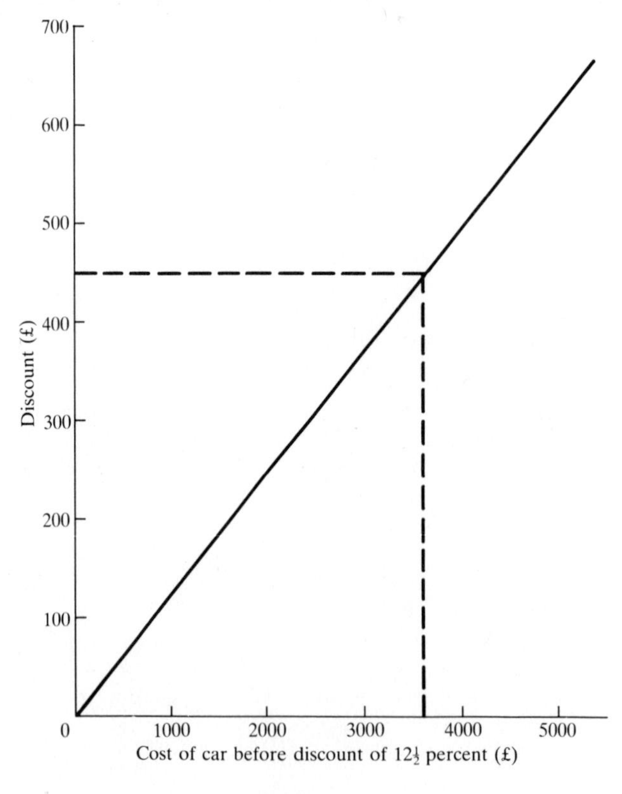

Fig. 7.1

If you need further help, turn to the extra work at the end of this section.

● Problems

1. A shop has the following notice in its window: '15% off all records and cassettes'. Mary sees a record priced at £4.60. How much does she have to pay to buy the record?

2. Mr Brown received the following bill.

Invoice No. 1124	W J Brookes & Son Ltd The Maze High Street Bushtown
	7.5.82
1 year's TV rental	£160.00
VAT at 15%	24.00
	£184.00

10% discount if paid within 7 days

How much should Mr Brown pay if he pays within 7 days?

3. Mary wants to buy a record. Jones Bros. has a sale with 15 per cent off marked prices and the record is marked at £5.00. Hill and Son also have a sale but with 12½ per cent off marked prices and the same record is priced at £4.80. From which shop should Mary buy the record?

4. Roger Wall bought a car and paid £400.00 to insure it. After one year the insurance company gave him a 20 per cent discount because he hadn't had an accident. How much does he now have to pay?

5. David Rogers asked two insurance companies how much it would cost to insure his car. The Motorists' Insurance Co quoted £450 less 60 per cent 'no claims' bonus and the Car Insurance Co quoted £500 less 65 per cent 'no claims' bonus. Which is the better quote?

6. A hotel charges £120 per week, but during January, February and March it offers a 15 per cent discount and during October, November, and December it offers a 12 per cent discount. Mr Johnson had a week's holiday in October and another week's holiday in February. What was the total cost of the two holidays?

7. A hotel charges £140 per week per adult with a 40 per cent reduction for children under 10 years of age. Mr and Mrs Grey have three children under 10 years of age and book a week's holiday at the hotel. What is the total cost of the week's holiday?

8. Mr and Mrs Simmons have two children, one aged 7 years and one aged 12 years, and they wish to spend a fortnight's holiday at The Villa Hotel. The Villa Hotel charges £110 per week per person with a 40 per cent reduction for children under 10 years and a 30 per cent reduction for children between 10 and 15 years. How much will it cost them?

9. Mr Hanson has a magazine which costs 80 p per month. If he pays one year's subscription in advance he gets a 15 per cent reduction. If he pays two subscriptions of six months each, he gets a 12½ per cent reduction. Compare the cost of one year's magazines if he pays monthly, half-yearly, or yearly.

EXTRA WORK

● Example A

In a sale all items are reduced by 12½ per cent. If a washing machine is marked at £320, what is the sale price?

Method 1

10% of £320	=	£32.00
2½% = ¼ of 10%	=	£ 8.00
Discount		£40.00
Marked price		£320.00
Discount		40.00
Sale price		£280.00

Method 2

Discount = £320 X 12.5/100 = £40

Marked price	£320.00
Discount	40.00
Sale price	£280.00

Method 3 Sale price = £320 X 87.5/100 = £280.

● Exercises A

1. A dress is marked at £40. But in a sale it is subject to 12½ per cent off. What is its sale price?

Method 1

10% of £40	=	£
2½% = ¼ of 10%	=	£______
Discount	=	£______
Marked price		£
Discount		£______
Sale price		£______

Method 2

Discount = £40 X 12.5/100 = £

Marked price	£
Discount	£______
Sale price	£______

Method 3 Sale price = £40 X 87.5/100 = £______

2. In a sale a settee is subject to 15 per cent off. If the original cost was £550, what was the sale price?

Method 1

10% of £	=	£
5% of £	=	£______
Discount	=	£______
Marked price		£
Discount		£______
Sale price		£______

Method 2

Discount = £ X 15/100 = £

Marked price	£
Discount	£______
Sale price	£______

Method 3 Sale price = £______ X 85/100 = £______

In questions 3–7 each price is subject to a 15 per cent reduction. What are the reduced prices?

3. £2.60; 4. £16.40; 5. £240; 6. £383; 7. £458.

In questions 8–12 each price is subject to a 12½ per cent reduction. What are the reduced prices?

8. £3.60; 9. £44.80; 10. £280; 11. £456; 12. £127.20.

13. Reduce £48 by 20 per cent; 14. Reduce £56 by 25 per cent;

15. Reduce £120 by 12½ per cent; 16. Reduce £324 by 33 1/3 per cent.

● Example B

The graph in Fig. 7.2 shows how much a 15 per cent discount is worth on prices up to £100. Thus, to reduce £60 by 15 per cent,

From Fig. 7.2, discount = £9
Reduced price = £(60–9)
= £51.

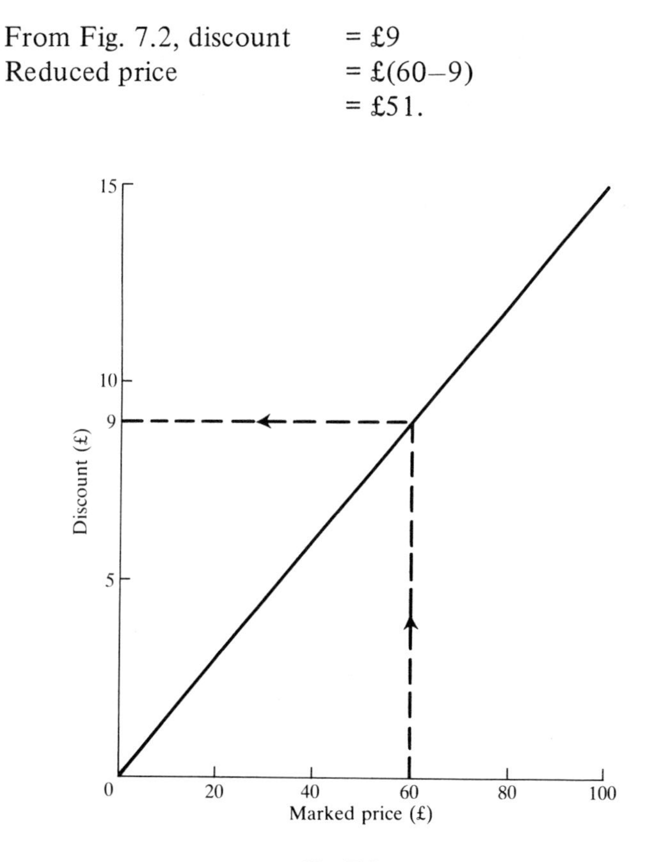

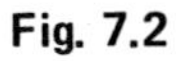
Fig. 7.2

● Exercises B

1. In a sale a set of saucepans are reduced from £20 by 15 per cent. What is the sale price?
From Fig. 7.2 reduction = 15% of £20 = ; sale price = £(–) = £

Use Fig. 7.2 to reduce the following by 15 per cent.

2. £30; 3. £50; 4. £90; 5. £80; 6. £70; 7. £15; 8. £65; 9. £95; 10. £85.

11. Draw a graph to show a discount of 20 per cent on amounts up to £100.

Use your graph to reduce the following by 20 per cent.

(a) £20; (b) £35; (c) £55; (d) £70; (e) £95.

7.2. Discount groceries

According to many advertisements, shopping for the family at a big supermarket seems to be *much* cheaper. But there is a catch – how far do you have to travel? And how much does it cost to get there? How far is it worth travelling to make a saving?

For example, a housewife living in Downend has to travel 7 miles to the hypermarket at Yate. Her car travels approximately 40 miles on each gallon of petrol and the cost of petrol is £1.60 per gallon.

If 40 miles cost 160 p,

$$1 \text{ mile costs } 160/40 = 4\text{p } (£0.04).$$

Thus the round trip of 14 miles costs

$$14 \times 4 \text{ p} = 56 \text{ p } (£0.56)$$

so that it costs her £0.56 to make her visit.

● Problems

1. Copy and complete Table 7.1 and check that the total saving *is* greater than 56 p.

Table 7.1

Item	Price at local shop (£)	Price at supermarket (£)	Saving (£)
1 packet *Persil Automatic*	1.85		0.16
800 g butter	1.68	1.60	
4 packets digestive biscuits	1.40	1.26	
800 g cheese	2.62	2.38	
Total saving			

2. (i) Write out a shopping list of groceries for one week to feed your family. Find the cost of each item at *your* local shops and work out the total cost.

(ii) Get a price list from your nearest larger supermarket and work out the total cost of buying your shopping list there instead (*not* counting travel).

(iii) Work out the distance from your home to the large supermarket, and how much it costs you to drive there. Add this to the cost of the groceries. Do you make a saving overall?

3. Consider the following shopping list and article prices.

	'local' grocers price	*large supermarket price*
2 pkts cornflakes	98p	92p
400 g butter	83p	80p
4 large loaves	£1.80	£1.72
2 dozen eggs	£1.45	£1.32
400 g bacon	£1.56	£1.38
200 g instant coffee	£1.81	£1.69
100 g tea	45p	41p
4 toilet rolls	76p	65p
1 frozen chicken	£4.05	£3.72

Work out the total cost of this shopping list: (a) at a local shop; (b) at a large supermarket.

There is a shop belonging to a chain of grocers in each of the towns marked on the map in Fig. 7.3. Work out the *distance* of each town, along the roads marked, to the hypermarket at Yate. Write the distance in the table below. Work out the cost of each journey and enter it in the table. In the last column, write Yes or No according to whether the saving at the hypermarket is bigger or smaller than the cost of the journey.

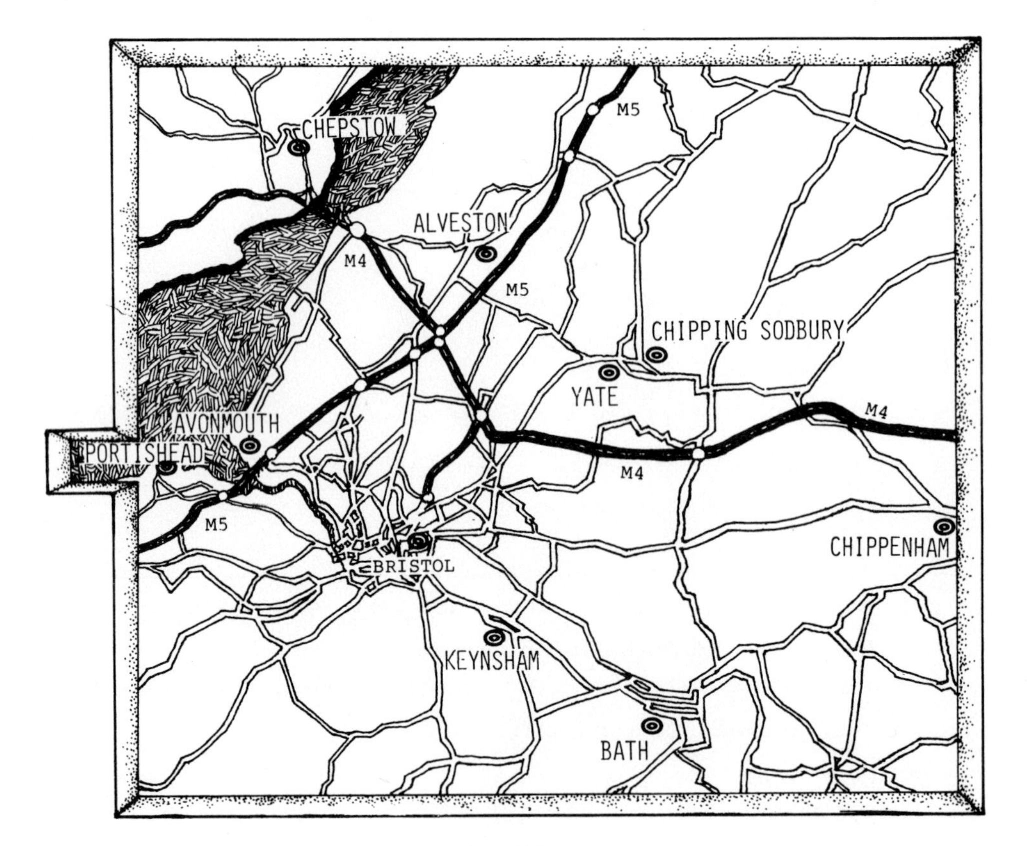

Fig. 7.3

Place	Distance to Yate (mi)	Distance there and back (mi)	Cost of journey (£) (at £0.04 per mile)	Worth the journey
Alveston				
Chepstow				
Chipping Sodbury	2	4	0.16	Yes
Avonmouth				
Portishead				
Keynesham				
Bath				
Chippenham				

4. The hypermarket starts to sell discount petrol at £1.52 per gallon instead of £1.60. Work out the new cost per mile of driving a car which goes 40 miles to one gallon. Work out how much it now costs (rounded to the nearest penny) to drive to Yate from each of the towns on the map. Does the cheap petrol alter any of your ‘No’ answers to ‘Yes’?

7.3. A good deal?

Alan Benn asks two garages for quotations for repairing his motor cycle and receives the quotations shown in Fig. 7.4. Which is the better deal?

In order to compare the estimates both need to include VAT which is charged at the rate of 15 per cent. The VAT can be found in any one of the following ways.

Brian J Last Ltd
12 Bridge Street
Stourbridge

6.5.82

Mr A Benn
The Woodlands
Woodside
Stourbridge

To repairing and road testing motor cycle £75

Hoping to receive your instructions

BJ Last

pp Brian J Last Ltd

All prices subject to VAT at standard rate

F. Wall & Sons Ltd
35 The Wayside
Stourbridge

7.5.82

Mr A Benn
The Woodlands
Stourbridge

To repairing and road testing motor cycle £86 inc VAT

Hoping to receive your instructions

F. Wall

pp F. Wall & Sons Ltd

Fig. 7.4

METHOD 1

Brian J. Last's quotation is £75 excluding VAT. Treating £75 as 100 per cent it is possible to work out 10 per cent and 5 per cent which can then be added to find the VAT.

	10% of £75	= £7.50
	5% of £75	= £3.75
VAT =	15% of £75	= £11.25

Brian J. Last's quotation exclusive of VAT is £75.00 + £11.25 = £86.25.
Thus the second quote is 25 p better than the first.

METHOD 2

The VAT can also be found by finding 15per cent of £75 and adding it to the £75 quote.

VAT on £75 = £75 X 15/100 = £11.25
True cost of the first quote = £75.00
11.25
£86.25.

METHOD 3

The true cost of the quote can be found by treating the £75 quote as 100 per cent and evaluating the true cost as a percentage.

True cost percentage = (100 + 15) per cent = 115 per cent.
True cost = £75 X 115/100
= £86.25.

METHOD 4. Graphical solution

VAT of 15 per cent means that 15 p on every £1 or £15 on £100 is added to the quote. Thus,

VAT on £0 = £0
VAT on £100 = £15.

From this information we can draw a graph (Fig. 7.5) to show the VAT on amounts from £0 to £100. Using Fig. 7.5,

VAT on £75 = 11.25
Total cost = £75 + £11.25
= £86.25.

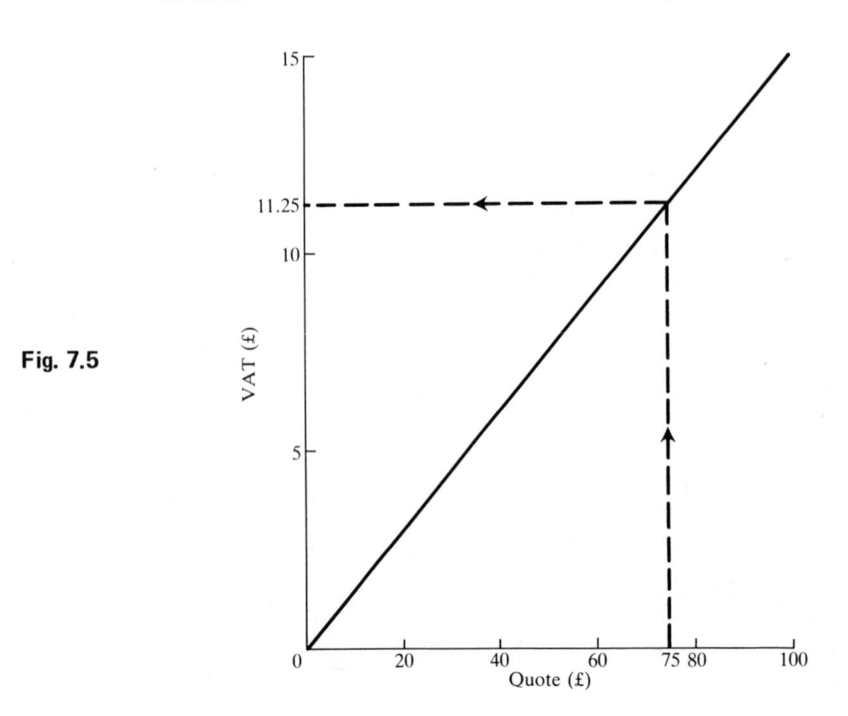

Fig. 7.5

Note that since the difference between the two quotes is so small, Alan Benn will probably make further enquiries about the two firms to see which is likely to do the better job.

If you need any further help before tackling the following problems, then turn to the extra work at the end of this section.

● Problems

1. Barbara Clarke has a hi-fi system which requires repairs. She obtains the quotes in Fig. 7.6. Which is the better quote?

Music For You Ltd
High Street
Kinver

8.5.82

Miss B Clarke
12 Brook Street
Kinver

To repairing Hi-Fi £40

Hoping for your instructions

D Allen

pp Music For You Ltd

All prices subject to VAT at standard rate

The Hi-Fi Centre Ltd
Bridge Street
Stourbridge

9.5.82

Miss B Clarke
12 Brook Street
Kinver

To repairing Hi-Fi £47 inc VAT

Hoping to receive your instructions

B. Higgs

pp The Hi-Fi Centre Ltd

Fig. 7.6

2. Joan Grey has a washing machine which needs repairing. She has two quotes. W. Mansell & Son quote £90 plus VAT and Hill's Washing Machine Co quote £102 including VAT. Who should she employ?

3. Mr Allen wants his house decorated. He asks two firms for estimates. Brian J. Painter Ltd quotes £750.00 plus VAT and F. Wall & Sons Ltd quotes £860.00 including VAT. Which is the better quote?

4. Mr Williams books a holiday in Spain and is quoted a cost of £230. However, when it is time to go, the travel agent puts a 15 per cent surcharge on the cost. What does the holiday now cost?

5. Mr and Mrs Brown have single glazing in their kitchen. They decide to have a new window. If the new window is single glazed it is subject to VAT, but if it is double glazed there is no VAT. The quotation gives £270 plus VAT for single glazing and £320 for double glazing. Compare the true costs.

6. Mr Bloxham wishes to have his drive tarmacked. He has three quotes. New Drives Ltd quote £600 including VAT, Roads & Bridges quote £620 including VAT, and Paths & Drives quote £550 plus VAT. Which is the cheapest quote?

7. Mr and Mrs Harrison buy an old cottage which needs a lot of work doing to it. They receive three quotes for rewiring the house. The local electricity board quote £750 including VAT. J.W. Electrics Ltd quote £650 plus VAT and F. Hill & Sons quote £680 plus VAT. Which quote is the cheapest?

8. Last year Mr Kyte paid £560 in rates. This year the local council have increased the rates by 20 per cent. How much does Mr Kyte now have to pay?

9. A shopkeeper buys televisions at £330 and wishes to make 25 per cent profit on each set. For how much should the television sets be sold?

10. A firm of greengrocers buy £118000 worth of goods during a year. Their accountant tells them that their profit was 20 per cent. How much was their total takings for the year?

11. A publican makes a profit of about 22 per cent on beer. If he pays £60 for an 18-gallon barrel, how much should he charge per pint?

EXTRA WORK

● Example A

A decorator quotes £80 excluding VAT, to decorate a room. What is the true cost?

Method 1

10% of £80 = £8.00
5% of £80 = £4.00

VAT = £12.00

True cost = £(80 + 12) = £92

Method 2

VAT = £80 × 15/100 = £12

True cost = £(80 + 12) = £92

Method 3

Cost including VAT = £80 × 115/100 = £92

● Exercise A

1. A black and white television is advertised at £120 plus VAT. What is the true cost?

Method 1

10% of £120 = £
5% of £120 = £______
VAT = £______
True cost = £(120 +) = £______

Method 2

VAT = £120 × 15/100 = £

True cost = (£120 +) = £______

Method 3

Cost including VAT = £120 × 115/100 = £______

2. A colour television set is advertised at £350 plus VAT. What is the true cost?

Method 1

10% of £ = £

5% of £ = £17.50

VAT = £______

True cost = £(+) = £______

Method 2

VAT = £ × 15/100 = £

True cost = £(+) = £______

Method 3

Cost including VAT = £ × 115/100 = £______

The following are quotes excluding VAT. Find the true costs.

3. £40; 4. £140; 5. £180; 6. £260; 7. £440; 8. £30; 9. £90; 10. £150; 11. £270; 12. £310.

● Example B

A sewing machine is advertised at £140 plus VAT. What is the true cost? The graph in Fig. 7.7 can be used with Method 4 to find the VAT on amounts from £0 to £500. From Fig 7.7, VAT on £140 = £21. True cost = £(140 + 21) = £161.

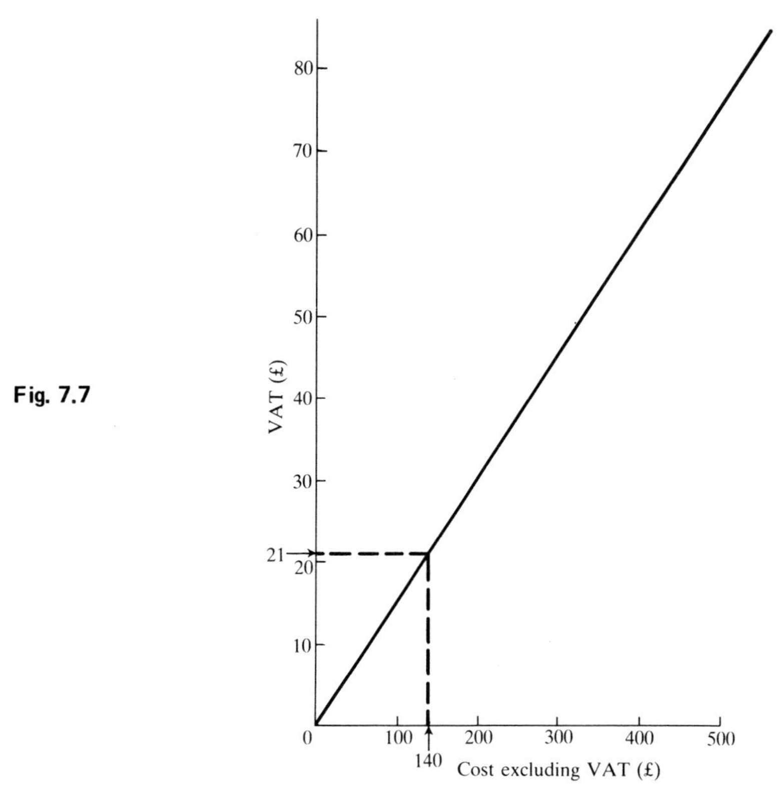

Fig. 7.7

● Exercise B

1. A portable television is advertised at £60 plus VAT. What is the true cost? From Fig. 7.7, VAT on £60 = £ ; True cost = £(60 +) = £______

The following are quotes excluding VAT. Use Fig. 7.7 to find the true costs.

2. £120; 3. £180; 4. £260; 5. £380; 6. £460; 7. £90; 8. £210; 9. £290; 10. £330; 11. £490.

7.4. Buying by instalments

When buying articles which cost more than a few pounds it is often possible to pay by instalments. Firstly, a deposit is paid by cash. Then a small amount is paid each month (or sometimes each week) over a period of time. This period might be a year or several years. We call this method of payment *hire purchase*. Unfortunately, it tends to cost more if you pay by instalments than if you pay cash at the beginning. According to the advertisement at the beginning of this section the local bicycle shop is selling a second-hand motorcycle for £250. You can either pay cash or pay by instalments after first paying a deposit of £50. There are 12 instalments each of £18.75. But how much extra do you pay?

The first step is to work out how much is borrowed from the shopkeeper. This is calculated as the difference between the cost of the motorcycle and the deposit.

Step 1: Amount borrowed = cost – deposit = £250–£50 = £200.

To calculate the extra payment by instalments we have to work out the total amount repaid to the shopkeeper and take off the amount borrowed.

Step 2: Total repayments = 12 × £18.75 = £225
Step 3: Extra amount paid = £225 – £200 = £25.

Thus, if you bought the bicycle by hire purchase, it would cost £25 more than by paying cash.

The extra work at the end of this section contains an example which is worked in more detail. If you cannot do the problems below, then have a look at this example and the exercises which follow on.

● Problems

1. The motorcycle costs £300 and you pay £50 deposit and 12 instalments of £27. How much extra would you pay by instalments?

2. The motorcycle costs £250 and you pay £100 deposit and 12 instalments of £14. How much would you have saved if you had paid by cash?

3. The motorcycle costs £500 and you pay £125 deposit and 12 instalments of £35.20. How much extra would you pay by instalments?

4. The motorcycle costs £500 and you pay £125 deposit. Now you pay 24 instalments of £19.70. How much extra would you pay by instalments?

5. Two motorcycles cost £400 and the deposit is £100. You pay 12 instalments of £28.13. Your friend pays 24 instalments of £15.63. Whose motorcycle costs more?

6. You are buying a television video game for £55 and having paid £15 deposit you may pay the rest over 24 months paying instalments of £2. Your friend decides to buy the same game but pays 12 instalments of £4. Who pays most interest?

EXTRA WORK

● Example

A motorcycle costs £450 but you can pay £100 deposit and 12 instalments of £35. How much extra do you pay by instalments?

Question	*Calculation*
1. How much do you borrow?	£450 – £100 = £350
2. What is the total repayment?	12 (no. of payments) × £35 (instalments) = £420
3. How much extra do you pay?	£420 (amount paid) – £350 (amount borrowed) = £70.

Thus the extra amount the motorcycle costs if you pay by instalments is £70.

Now try these exercises.

● Exercises

1. The motorcycle costs £250 and you pay £100 and 12 instalments of £15. How much extra do you pay by instalments?

Question	*Calculation*
1. How much do you borrow	£250 – £100 =
2. What is the total repayment?	12 × £15 =
3. How much extra do you pay?	– =

2. The motorcycle costs £350 and you pay £50 deposit and 12 instalments of £29. How much extra do you pay by instalments?

Question	*Calculation*
1. How much do you borrow?	
2. What is the total repayment?	
3. How much extra do you pay?	

These questions show the steps that you need to take to solve the problems.

7.5. The cost of advertising

Write your advertisement here (*one word per box*)					
					■
					64p
					96p
					£1 – 28

Note : A telephone number or a price counts as one word.
A time (e.g. 5 p.m.) or an abbreviation counts as one word

Fig. 7.8

Write your advertisement here (*one word per box*)					
LADIES	BICYCLE	for	sale.	Good	■
condition	cost	£95	new	will	64p
accept	£50	or	nearest	offer	96p
Tel.	———				£1 – 28

Note : A telephone number or a price counts as one word.
A time (e.g. 5 p.m.) or an abbreviation counts as one word

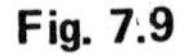

Fig. 7.9

It is very easy for private advertisers to sell items in their local newspapers. One local paper offers advertisements starting at 64 p (for up to 10 words). You simply fill in a coupon like the one in Fig. 7.8 and send it, with the correct money, to the newspaper. Or, you can call at the newspaper office and they will help you word your advert.

Fig. 7.9 shows a newspaper advert in its finished form and as sent in on a coupon like that in Fig. 7.8. The cost would be £1.28 for this advert.

● Problems

1. How much would each of the following advertisements cost?

(a) HOOVER JUNIOR vacuum cleaner. Like new, 6 months guarantee, free delivery. £29.99. – Tel:

(b) MODERN electronic coin operated piano, complete with 70 eight track tapes. £150 ono. Tel:

(c) SINGER ELECTRIC sewing machine with attachments. £29. – Tel: after 5 p.m.

(d) TEAC A33405 four-channel tape recorder. Teac AN 80 Dolby system. Foiling console. £750 ono.

(e) IBM ELECTRIC TYPEWRITER (not portable size), perfect condition. £35. – Tel after 6 p.m.

(g) WANTED FOR CASH, old bedroom furniture, sideboards, tables and chairs, solid wood pre–1960 etc. Full house contents. Tel:

(g) VICTORIANAS ANTIQUES and collectors fair. Forum Theatre, Cannock, Saturday, February 20, 10.00 a.m.– 5.00 p.m. Signposted. Refreshments.

(h) WILTON CARPET, 100 per cent wool, 10ft x 9ft. Rust / cream / gold. £70 ono. – Tel: after 6 p.m.

(i) TEAK GAS FIRE, with back boiler, suitable six to eight radiators, brand new. Tel: Stratford.

2. If you wish to spread out your advertisement to give it more impact, you can pay by the amount of space you use. The unit of measurement used is a column-centimetre. One local paper charges classified advertisements at £2 per column-centimetre.

(a) What is the cost of an advertisement that is one column wide and one cm long?

(b) What is the cost of an advertisement that is one column wide and 2 cm long?

(c) What is the cost of an advertisement that is one column wide and 10 cm long?

3. If an advertisement is more than one column wide, you multiply the number of columns by the number of cm to get the number of column-centimetres. Find the number of column-centimetres in each of the following adverts:

(a) 2 columns wide, 3 cm long;

(b) 3 columns wide, 8 cm long;

(c) 2 columns wide, 5 cm long.

4. Copy and complete the table below to calculate the cost of each of the advertisements in Fig. 7.10.

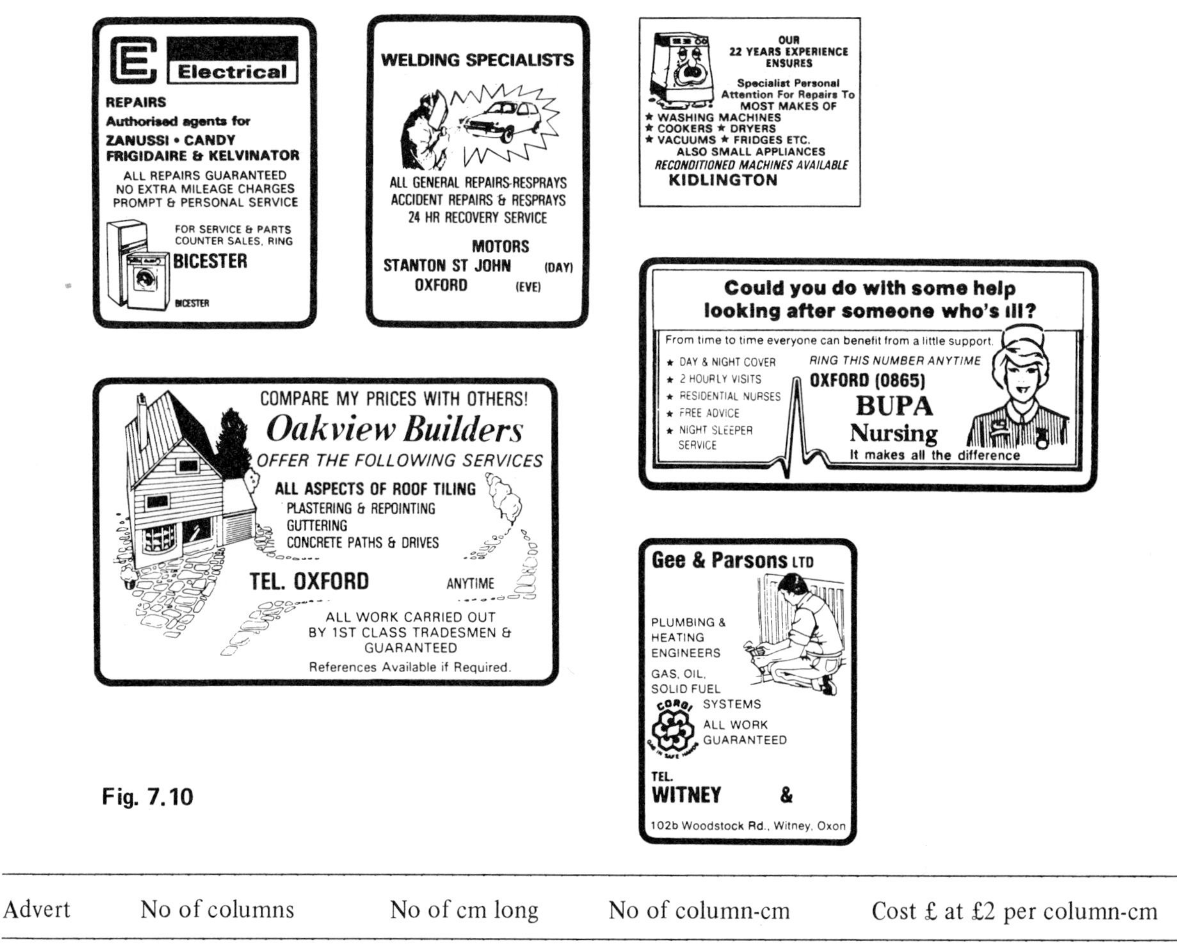

Fig. 7.10

Advert	No of columns	No of cm long	No of column-cm	Cost £ at £2 per column-cm
(a)	1	8	8	
(b)				
(c)				
(d)				
(e)				
(f)				

NB A client is generally charged for a little more than the length of the advert to allow for the gaps between the advertisements. The length is always measured as a whole number of cm.

● Related problems

A. Block advertisements

Advertising is expensive. It costs £10 000 for a full page advertisement in a national newspaper. Advertisements are usually smaller than this. They can be many sizes. To make costing simpler the page is split into eight columns and each column is usually split into 20 blocks (Fig. 7.11).

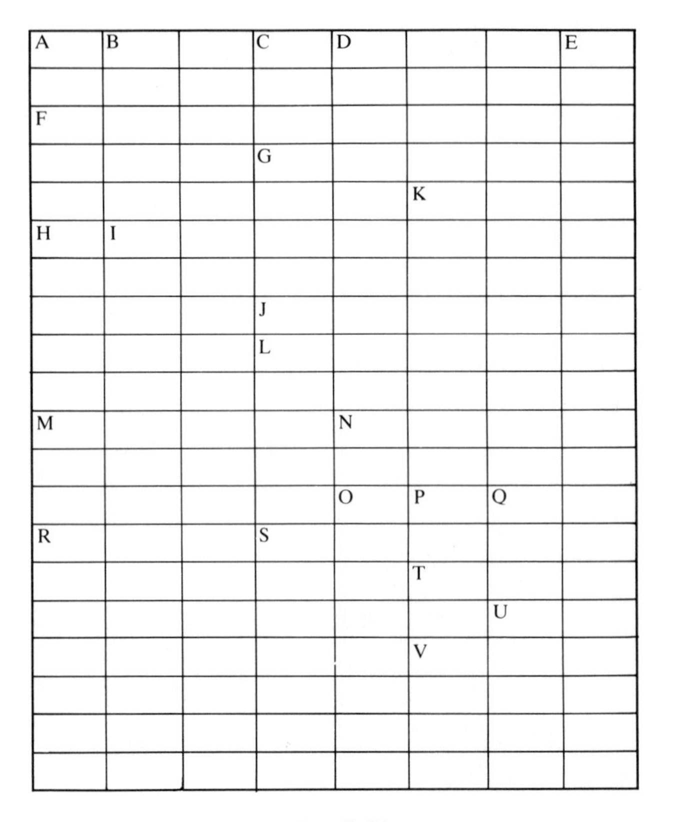

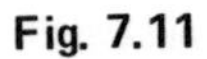
Fig. 7.11

1. Calculate the cost of advertisements of the following sizes:

   (a) Half page;
   (b) Quarter page;
   (c) One full column;
   (d) One half column;
   (e) A double column.

2.(a) In a page like the one in Fig. 7.11, how many blocks are there?
   (b) What is the cost of one block?
   (c) Calculate how much it would cost for each of the advertisements A to V shown on the newspaper page in Fig. 7.11.

B. TV advertising

Programmes on independent television are paid for by the money raised from television advertising. The advertising rates vary through the day, being highest during the peak period when most people will be watching television.

1. If a television company screens programmes according to the following timetable, calculate the number of viewing hours each day and then the total number of viewing hours per week.

Day	Start	Finish	Viewing times in hours
Monday	11 a.m.	11 p.m.	
Tuesday	11 a.m.	midnight	
Wednesday	10.30 a.m.	11.30 p.m.	
Thursday	11 a.m.	11.30 p.m.	
Friday	10 a.m.	midnight	
Saturday	9.30 a.m.	midnight	
Sunday	10.00 a.m.	11 p.m.	
	Total viewing time per week =		

2. Nine minutes of every hour's viewing time are allowed for advertising. How many minutes advertising are there in one week?

3. If the average rate of advertising is £8000 per minute, how much money is raised in one week?

4. How much money is raised in one year (52 weeks)?

5. If the TV company charged an average rate of £9000 per minute, how much would it raise in one year?

6. A new commercial TV company is bidding for a regional TV franchise. Amongst the details it submits to the Independent Broadcasting Authority is the following proposed schedule of programmes.

Monday	11 a.m.	11.30 p.m.
Tuesday	11 a.m.	midnight
Wednesday	10 a.m.	11.30 p.m.
Thursday/Friday	10.30 a.m.	00.30 a.m.
Friday/Saturday	11 a.m.	1.00 a.m.
Saturday/Sunday	9 a.m.	1.30 a.m.
Sunday	10 a.m.	11 p.m.

Nine minutes of every hour's viewing time are allowed for advertising. If this new company charges an average rate of £8500 per minute's advertising how much would it raise in: (a) 1 week; (b) 1 year?

8 Odds and ends

The subject matter of the earlier chapters in this book involve problems which at some time may occur in everyone's life. For example, planning a party is a real activity for many students, as is the desire to move into a flat of their own. These problems involve the use of simple mathematics in their solution and show the importance of grasping basic techniques.

In this final chapter we have gathered together 10 problems which don't fit directly into the subject areas of the earlier chapters. They are diverse in nature and, although only using simple mathematical ideas, we hope that your students will find these problems both interesting and intriguing.

8.1. 20 questions

Guessing the identity of an unknown object by asking 20 questions each of which can only be answered 'Yes' or 'No' was a popular radio programme. We have compiled 20 questions of a different nature. Each question is answered A, B, C, or D. Choose which you think is most appropriate. Work quickly through the 20 questions – you have 5 minutes. When you have done this, calculate the answers carefully. Some of the results may surprise you!

	A	B	C	D
1. What is your height in inches?	43–59	60–74	75–89	90–100
2. What is your height in cm?	70–110	110–150	150–190	190–230
3. What is your weight in lbs?	84–104	105–125	126–146	147–167
4. What is your weight in kg?	32–48	49–59	60–70	71–81
5. What is the diameter of a 10 p coin in cm?	1.5–1.9	2.0–2.4	2.5–2.9	3.0–3.4
6. How long is your left foot in cm?	12–20	21–29	30–38	39–47
7. How many bags of crisps can be made from one ½ lb potatoes	4–6	7–9	10–12	13–15
8. How many litres are equivalent to 5 gallons?	22–27	28–33	34–39	40–45
9. How many ml make one pint?	550–600	600–650	650–700	700–750
10. How many g make 1 oz?	15–18	19–22	23–26	27–30
11. What would be the cost of 1 pint of milk (20p) taken every day for 1 year in pounds?	43–53	54–64	65–75	76–86
12. What is the thickness of a sheet of writing paper in mm?	0.05–0.1	0.11–0.16	0.17–0.22	0.01–0.04
13. How many hours are spent in school in one year, if you have full attendance?	300–500	500–700	700–900	900–1100
14. What is your age in days?	5700–6000	6000–6300	6300–6600	6900–7200

15. What is the age in days of Christianity	25 000–75 000	125 000–250 000	1 000 000–2 000 000	700 000–900 000
16. What is the volume of the classroom in cubic metres?	20–120	120–220	220–320	320–420
17. Forty miles per h is the same as ____metres per second	8–12	12–16	16–20	20–24
18. The distance from the earth to the moon is about ¼ million miles. How many kilometres is this?	150 000–250 000	250 000–350 000	350 000–450 000	450 000–550 000
19. Forty miles to the gallon is equivalent to ____km per litre?	1–10	10–20	20–30	30–40
20. The distance from John O'Groats to Lands End is about 800 miles. How many metres is this?	300 000–600 000	700 000–1 000 000	1 100 000–1 500 000	3 000 000–5 000 000

8.2. Birthdays

Monday's child is fair of face
Tuesday's child is full of grace
Wednesday's child is full of woe
Thursday's child has far to go
Friday's child is loving and giving
Saturday's child works hard for his living
And the child that is born on the Sabbath day
is bonny and blithe and good and gay.

Have you ever wondered on what day of the week you were born?

While it is usual to remember the date of your birthday it is quite likely that your parents have forgotten the actual day of the week on which you were born. It is possible to work backwards, but having leap years every four years makes this difficult. There is a method which will work out the day of the week for any birthday this century.

We will show you how the method works and then give you some problems. You can check your answers and then work out on what day of the week you were born.

If the number of the year is exactly divisible by 4, then it is a leap year and February has 29 days. Otherwise it has 28.

For example, if we take 10 August 1968, 1968 divides exactly by 4. It is a leap year. February has 29 days.

Step 1: Write down the year 1968   1968

Step 2: Add up the days

January	31
February	29
March	31
April	30
May	31
June	30
July	31
August	10
September	
October	
November	
December	
Number of days	223

Step 3: Take 1 off the year and divide by four. Ignore the remainder. Thus,

$$1968 - 1 = 1967$$

$$\frac{1967}{4} = 491 \text{ if we ignore the remainder.}$$

Step 4: Add the answer	491
to the year number	1968
and the number of days	223
	2682
Subtract	− 15
	2667

Step 5: Divide by 7. The remainder gives the day of the week.

Remainder = 0 – Saturday
1 – Sunday
2 – Monday
3 – Tuesday
4 – Wednesday
5 – Thursday
6 – Friday

Dividing 2667 by 7 gives a remainder of 0. Thus, if you were born on 10 August 1968, you were born on a Saturday.

The method is as follows.

(1) Let Y = year and D = number of days up to birthday.
(2) Find $\frac{Y-1}{4}$, ignoring the remainder.
(3) Work out $S = Y + D + \frac{(Y-1)}{4} - 15$.
(4) Divide S by 7 and use the key remainder.
0 – Saturday
1 – Sunday
2 – Monday
3 – Tuesday
4 – Wednesday
5 – Thursday
6 – Friday.

● Problems

1. Find the day of the week for the date 4 June 1972. (Don't forget that 1972 was a leap year, and so February had 29 days.)
2. Find the day of the week for the date 29 July 1981.
3. Find the day of the week on which you were born.
4. What day of the week was Margaret Thatcher born on?
Her date of birth is 13 October 1925.
5. What day of the week was Kevin Keegan born on? His date of birth is 14 February 1951.
6. On what day of the week did the Second World War start? The date was 3 September 1939.
7. What day of the week was the Princess of Wales born on? Her date of birth is 1 July 1961.

If you have difficulties finding the answers to these problems try using the worksheet which follows.

● Worksheet

Follow these instructions to find out the day of the week for any date this century.

1. Write down the year.
2. Add up the days
   January
   February
   March
   April
   May
   June
   July
   August
   September
   October
   November
   December
   ______
   Number of days ______
3. Take 1 off the year and divide by four. Ignore the remainder.
4. Add the answer
   to the year number
   and the number of days ______
   Subtract 15
   ______
5. Divide by 7 the remainder gives the day of the week

   Remainder = 0 – Saturday
   1 – Sunday
   2 – Monday
   3 – Tuesday
   4 – Wednesday
   5 – Thursday
   6 – Friday.

8.3. Queues

On a particular day the probability of someone entering a Post Office during each minute is ½ and, if there is at least one person in the queue, the probability of someone leaving the queue in any period of one minute is 1/3. What is the expected length of the queue after 20 minutes? This type of problem can be investigated by a simple experiment.

We can use a die to find out if anybody arrives during a period of one minute by saying that if 1, 2, or 3 is thrown then somebody arrives. If 4, 5, or 6 is thrown, then nobody arrives. Now throw the die again and use 5 and 6 to represent someone leaving the queue (assuming there was somebody in the queue at the start of that minute) and 1, 2, 3, and 4 to represent no one leaving. This corresponds to a 1/3 chance of leaving the queue.The number in the queue at the end of the minute can be found in this way. A suggested tabular layout is shown in Table 8.1. If several pupils do this to simulate the first 20 minutes, the average length of queue may be found. You may be interested in trying different probabilities and seeing what happens.

Table 8.1

Minute	No. in queue at start	First throw	No. joining queue	Second throw	No. leaving queue	No. in queue at end
1st	0	3	1	–	0	1
2nd	1	3	1	5	1	1
3rd	1	5	0	6	1	0
4th	0	4	0	–	0	0
5th	0	2	1	–	0	1
6th	1	3	1	1	0	2
7th	2	2	1	3	0	3
8th	3	4	0	1	0	3
9th	3	1	1	6	1	3
10th	3	4	0	3	0	3
11th	3	6	0	5	1	2
12th	2	1	1	4	0	3
13th	3	6	0	3	0	3
14th	3	1	1	4	0	4
15th	4	4	0	3	0	4
16th	4	2	1	6	1	4
17th	4	3	1	6	1	4
18th	4	3	1	4	0	5
19th	5	3	1	5	1	5
20th	5	5	0	2	0	5

Other probabilities can be obtained by using spinners or numbered cards. For example, if the two probabilities required are 1/3 and 1/4 two spinners could be used, one the shape of an equilateral triangle and one in the shape of a square with a cocktail stick or match through the centre of gravity (Fig. 8.1).

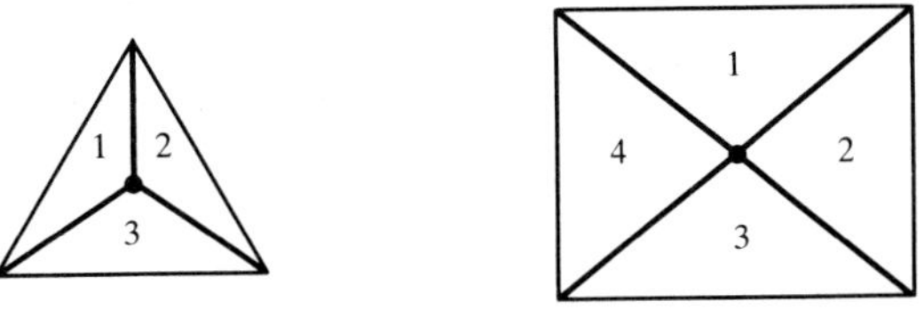

Fig. 8.1

An alternative method is to have a box with twelve cards numbered 1 to 12 in it. A probability of 1/3 can be represented by drawing a 1, 2, 3, or 4 and a probability of 1/4 by drawing a 1, 2, or 3.

Yet another method is to use a circular board, a pointer and interchangeable faces to represent different probabilities. A probability of 1/5 can be obtained by using the face as shown in Fig. 8.2.

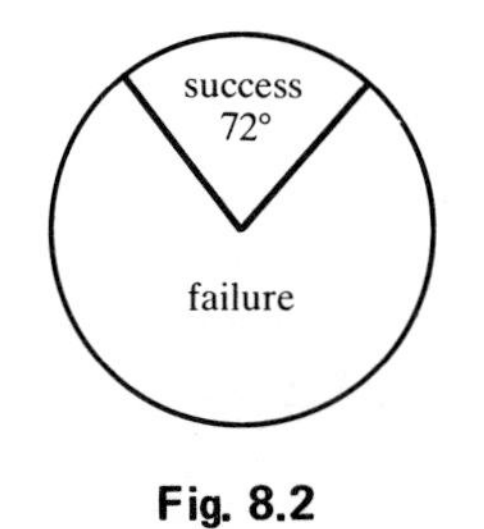

Fig. 8.2

If further help is needed, use the exercises at the end of this section.

Problems

1.

A garage owner notices that, if there are three or fewer cars waiting for petrol, the probability of another car joining the queue during a period of one minute is 1/2. However, if there are more than three cars in the queue this probability falls to 1/6. On average it takes 3 minutes to serve each customer. When would you expect the queue first to contain 4 cars? When would you expect it to drop back to 3 cars?

2. At a supermarket the probability of someone joining the queue in any period of one minute is 1/2 and the probability of someone leaving the queue in any period of one minute is 2/5 (assuming there is someone in the queue at the start of a minute.) What is the expected length of the queue after one hour? When would you expect the queue to first contain 3 people?

3. A supermarket finds that a customer arrives every 4 minutes on average and that it takes 5 minutes to serve each customer. How long will the queue be expected to be after 2 hours?

4. A shop has one assistant and the probability of a customer arriving during any period of one minute is 1/2 and the probability of a customer being served during any one minute is 1/3 (assuming there is someone in the queue at the start of the minute). How long would you expect the queue to be after 20 minutes? A second assistant now arrives, with the result that the probability of a customer leaving the queue becomes 2/3. When would you first expect the queue to disappear?

Exercises

1. Complete the following table in order to simulate Problem 1 for the first 20 minutes. A 1, 2, or 3 on the first throw represents someone leaving the queue, assuming there was someone in the queue at the start of the minute.

Minute	No in queue at start	First throw	No. joining queue	Second throw	No. leaving queue	No. in queue at end
1st	0	2	1	–	0	1
2nd	1	3		6	0	
3rd		3		4		
4th		5		1		
5th	2	1		3		
6th		1	1	4		
7th		3		2	1	
8th		4		5		
9th		6		3		
10th		1		2		4
11th		1		5		
12th		4	0	3	0	
13th		3		2		
14th		1		4		
15th	6	2	–	5		
16th		4	0	1		
17th		2		5		
18th		6		2	1	
19th		1		6		
20th		3		3		8

Now use your own die and draw a table to simulate a full 30-minute period.

2. Complete the following table in order to simulate problem 2 for the first 20 minutes. A 1, 2, or 3 on the first throw represents someone joining the queue if there are less than four cars in the queue. A 1 on the first throw represents someone joining the queue if there are four or more cars in the queue. A 1 or 2 on the second throw represents someone leaving the queue, assuming someone was in the queue at the start of the minute.

Minute	No. in queue at start	First throw	No. joining queue	Second throw	No. leaving queue	No. in queue at end
1st	0	1	1	–	0	1
2nd	1	3	1	5	0	
3rd		3		3		
4th		4		2		
5th		2		3		
6th		1		5		
7th	4	3	0	2		3
8th		2		6		
9th		1		1		
10th		5		2		
11th		3		5		
12th	4	5		6		
13th		3		4		
14th	4	2	0	1		
15th		3		2		
16th		4		1		
17th		3		6		3
18th		2		5		
19th	4	2		1		
20th		5		4		3

Now use your die and draw a table to simulate a full 30-minute period.

8.4. Who is supposed to be able to read this?

Fig. 8.3 gives two extracts from newspapers. Read them. Which one was the easier to understand? Can you think why this is so? How old do you think you need to be to understand what each passage is about?

OBSERVER REVIEW

A second main analytical strand would be about the fact that societies have sets of attitudes, partly specific to particular classes but also more widely shared, which usually work in contrasting clusters. So on the one hand there is that habitual unexpectancy in the English spirit, that process by which the belief in gradualness and tolerance and lack of showiness is always threatening to change into a damp, dog-in-the-manger, slightly bloody-minded, going-on-going-on; as in the current quite healthy rejection of the high gloss of some aspects of prosperous Europe which can degenerate into a commitment to squalor, to grime in the grain.

Contrast that with the devotion to unselfseeking voluntarism, or the enormous preoccupation with gentle spare-time activities, that poetry of amateurism which occupies millions in their spare time; or with the neighbourliness which is still habitually practised whenever one is out of semi-anonymous, mass, public situations. Or the sudden creative, generous, public gestures by which we surprise ourselves, such as the founding of the National Health Service or the Open University.

Then one would have to talk about the movements of the pendulum of a society's attitudes. Just now we are in a watchful, self-protective phase; one in

THE SUN

TONY NEAL is King of the Binmen. Our muck is his brass.

Neal, 37, was put in as managing director to see that Southend's streets were swept clean.

He says: "If your house has been missed you can phone the depot and a crew will go back and pick up your rubbish almost immediately."

Crews

Most of the staff are former council workers who now do an eight-hour day instead of the previous "task-and-finish" system.

Yard foreman Ron Dedman, 43, says: "When we worked for Southend Council, there was one job that two men used to do every Saturday.

"All it involved was checking the lights and indicators were working on all the vehicles. They were paid overtime for that. No one else could get a look-in."

Dustman Colin Todd, 42, is happy with his new employers in Southend.

Rise

As a ganger, or team boss, he now earns a basic £94 a week, and around £130 with overtime. A pay rise in April will bring his basic wage to £100.

Fig. 8.3

Many formulae have been devised to work out which books are suitable for readers of a certain age. The simplest of these is the Forecast formula, which gives the reading age (R years) of a book as

$$R = 25 - N/10$$

where N is the number of one-syllable words in a passage of 150 words.

In Table 8.2, we have worked out N for the *Observer* Review article. Here we have taken just the first 150 words of the article, ignoring numbers, and counting words like 'eight-hour' as two words.

Table 8.2

Sentence number	Number of one-syllable words
1	18
2	49
3 (unfinished)	24

Thus, the reading age of the *Observer* Review is

$$R = 25 - 91/10$$
$$15.9 \text{ years.}$$

Was your guess of the reading age anywhere near this?

● Problems

1. Following the method used above, work out the reading age R for the *Sun* article.
2. Read the following piece and work out its reading age.

One upon a time there was a mother pig who had three little pigs.

The three little pigs grew so big that their mother said to them, 'You are too big to live here any longer. You must go and build houses for yourself. But take care that the wolf does not catch you.'

The three pigs set off. 'We will take care that the wolf does not catch us,' they said.

Soon they met a man who was carrying some straw. 'Please will you give me some straw?' asked the first pig. 'I want to build a house for myself.'

'Yes,' said the man and he gave the first pig some straw.

Then the first pig built himself a house of straw. He was very pleased with his house. He said, 'Now the wolf won't catch me and eat me.'

'I shall build a stronger house than yours,' said the second little pig.

'I shall build a stronger house than yours, too,' said the third pig.

3. In the reading age formula

$$R = 25 - N/10,$$

what is the largest possible value of N? What then is the smallest possible value of R? Can you see the problem you've unearthed?

4. Think again about the two newspaper articles. Is it only the short words which make it easier to understand the *Sun* article?

How about the length of the sentences? Fill in the column headed 'words' in Table 8.3, with the number of words in each sentence of the *Sun* up to 150 words. Fill in the column for one-syllable words from the table you prepared in Problem 1 and then fill in the number of 2, 3, 4, or 5-syllable words, if any, in each sentence.

Now make a copy of Table 8.3 and fill it in for the *Observer* article (the number of sentences which contain 150 words will, of course, be different). This shows clearly the different lengths of sentences and the number of 'long' words in each passage. What is the average number of words per sentence for each article?

Table 8.3

Sentence	Words	Number of syllables				
		1	2	3	4	5
1						
2						
3						
4						
5						
6						
7						
8						
9						
10						
11						
12						
13 (unfinished)						
Total	150					

Calculate the average as W = total number of words/number of sentences, but only use *whole* sentences. Which paper usually uses longer sentences?

Make out a table like Table 8.3 for the children's story, too. What is the average sentence length for this story?

● Related problems

A. Another formula for calculating reading age (R) is the FOG formula (FOG stands for frequency of gobbledegook!). This takes into account the average length of the sentences and the frequency of long words.

$$R = 0.4\,(W + p)$$

where W = average number of words in a sentence and p = percentage of words with three or more syllables.

Calculate the percentage as

$$p = \frac{\text{number of words of 3 or more syllables}}{150} \times 100.$$

For the *Observer*, we saw earlier that

Number of sentences	= 2
Number of words in these	= 109
Number of 3-syllable words	= 10
Number of 4-syllable words	= 9
Number of 5-syllable words	= 7.

Thus, the average number of words = W = 109/2 = 54.50 and the percentage of long words = p = (10 + 5 + 2)/150 × 100 = 11.33.

We finally have

$$R = 0.4\,(54.50 + 11.33)$$
$$= 26.3 \text{ years}$$

as the reading age for the *Observer* Review.

Now *you* work out the reading ages of the *Sun* and the child's story. Do these agree with what you expected? Can you think of the advantages and disadvantages of this method?

B. A third formula takes into account the average sentence length and the *total* number of syllables in 100 words, not just the number of very short or very long words.

Use Table 8.3 to work out the total number of syllables, but remember you need 100 words and not 150.

Now work out the reading ages of the two newspapers from

$$R = 2.7971 + (0.0778 \times W) + (0.0455 \times S)$$

where W = the average number of words in sentence (use answer to Problem 4) and S = the total number of syllables in 100 words. Do these seem reasonable reading ages?

What about the child's story. What reading age is it meant for? What are the advantages and disadvantages of this method?

C. Rudolf Flesch produced a formula

$$\text{score} = 206.835 - (0.846 \times S) - (1.015 \times W)$$

where S = total number of syllables in 100 words and W = average number of words in a sentence. What is the score for the *Sun*?

This score is *not* a reading age – as you may have guessed! You need to use Table 8.4 to work out the reading age.

Table 8.4

Score	Reading age
Over 70	$5 + \frac{(150 - \text{score})}{10}$
Over 60	$5 + \frac{(110 - \text{score})}{5}$
Over 50	$5 + \frac{(93 - \text{score})}{3.33}$
Under 50	$5 + \frac{(140 - \text{score})}{6.66}$

What are the reading ages for the *Observer* and the children's story using this method? Do these seem reasonable?

A simpler version of the reading age is given by Flesch's nomogram, given in Fig. 8.4.

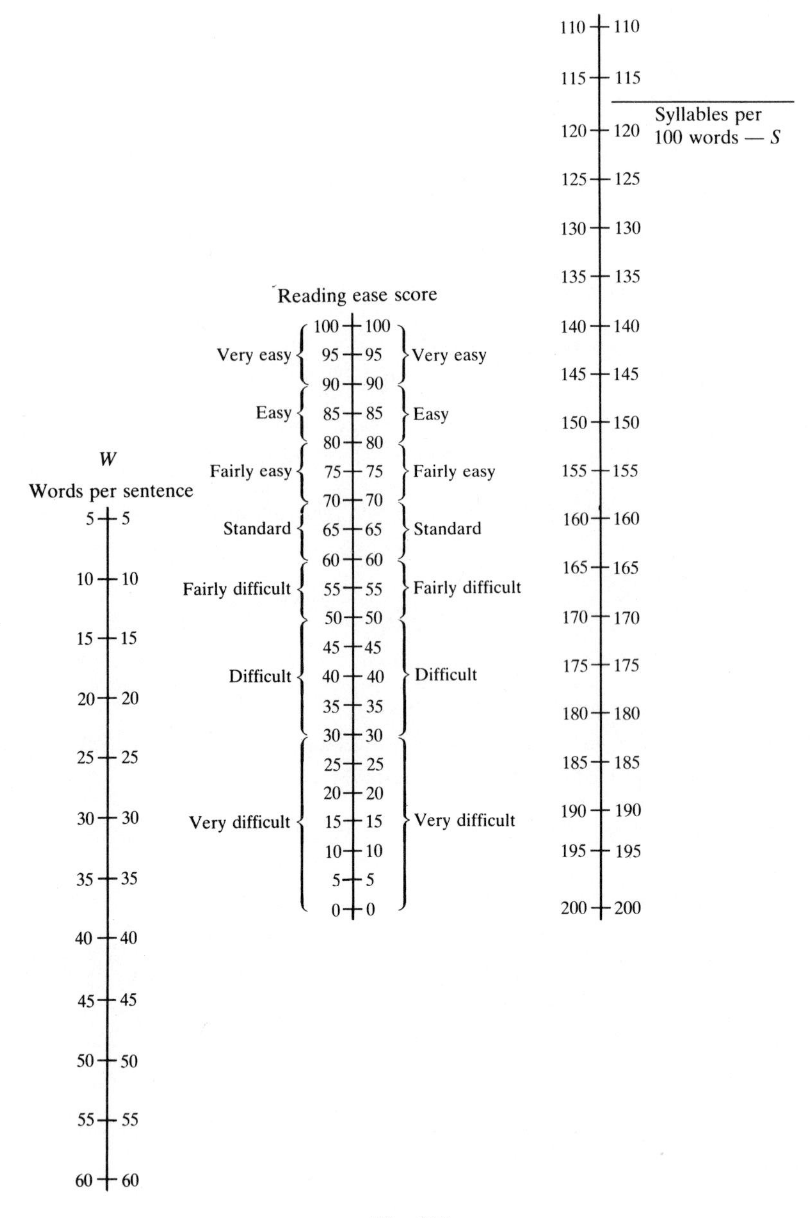

Fig. 8.4

You need a ruler to draw a line from the value of W in the left-hand column, to the value of S in the right-hand column; where the line crosses the centre column you can read the value of the score. Do this for both newspapers.

Now use Table 8.4 to work out the reading ages. Do your results agree with the values calculated using the Flesch formula for score at the beginning of this problem?

Can you work out the age for which the children's story is suitable, too?

D.(i) Guess the reading age of your favourite magazine and then use one or two of the methods to see how close your guess was.

(ii) Who is expected to be able to understand official forms? Use one of the tests on the 'small print' on a government form, hire purchase agreement, insurance policy, or similar document. Do you agree with the reading age suggested?

8.5. Plugging gaps

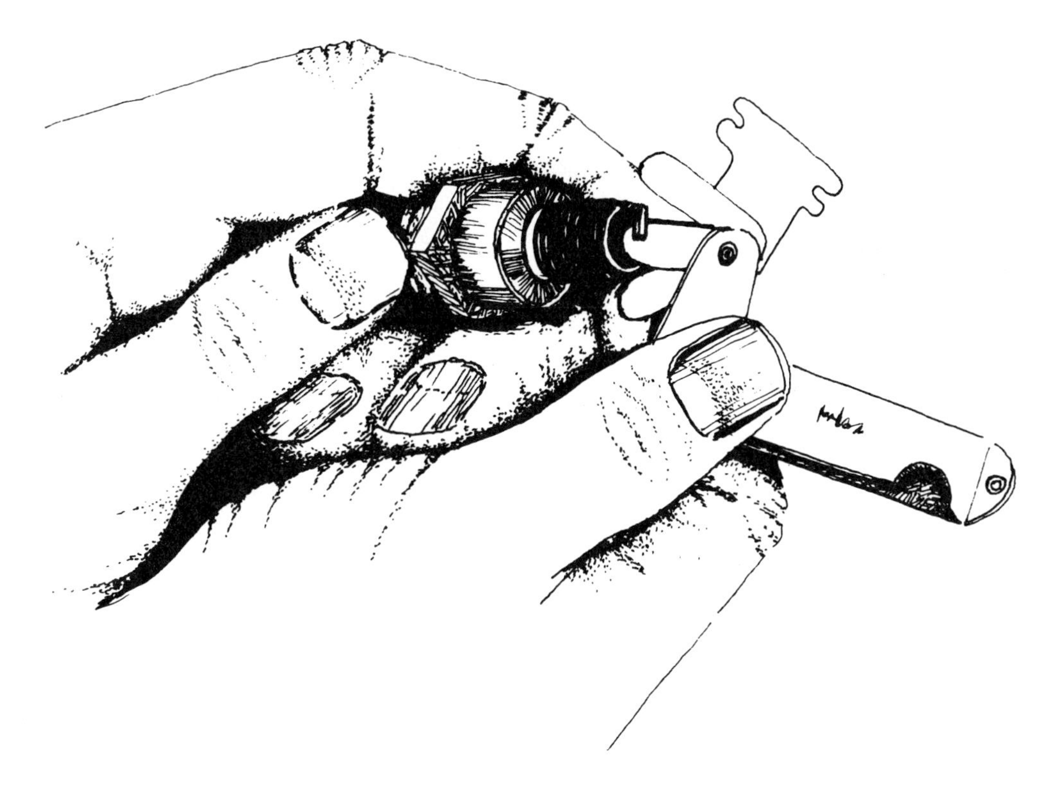

You may have seen the feeler gauges used to measure the gaps in spark-plugs.

They look rather like pen-knives with thin flexible blades. A very simple gauge has two blades which are of thicknesses 0.05 mm and 0.10 mm. With it you can measure gaps of 0.05 or 0.10 or 0.15 mm, using the blades singly or the two together. If you don't like these small decimal numbers you could think of the gaps as being 1, 2, and 3 units, a unit being 0.05 mm.

● Problems

1. With three blades, thicknesses 0.05, 0.10, 0.20 mm (1, 2, 4 units) you could measure seven widths:

0.05, 0.10, 0.15, 0.20, 0.25, 0.30, 0.35 mm (1, 2, 3, 4, 5, 6, 7 units).

Complete the table by ticking which blade or combination of blades must be used to measure each gap width.

Gap width (× 0.05 mm)	Blade thickness (× 0.05 mm) 1	2	4
1			
2			
3			
4			
5			
6			
7			

2. Now design a feeler-gauge with four blades. Find the 'best set' for measuring successive multiples of 0.05 mm. Make a table to show how to make up the 15 widths from 0.05 mm to 0.75 mm (from 1 to 15 units) as in Problem 1.

3. How about five blades? You need not stop there. See if you can find the following for a feeler-gauge with any number of blades:

(a) A rule for which blade-thicknesses to choose;
(b) A pattern in the table showing how to make up all the possible widths, in succession;
(c) A formula for the total number of widths you can measure.

● Related problems

A. Suppose you had these five coins.

1 p 2 p 2 p 5 p 10 p

How many different sums of money could you make? Think about it before you read on. Did you find 20? List them all if you're not sure.

1 p
2 p
3 p = 1 p + 2 p
4 p = 2 p + 2 p

and so on. Or you could make another table.

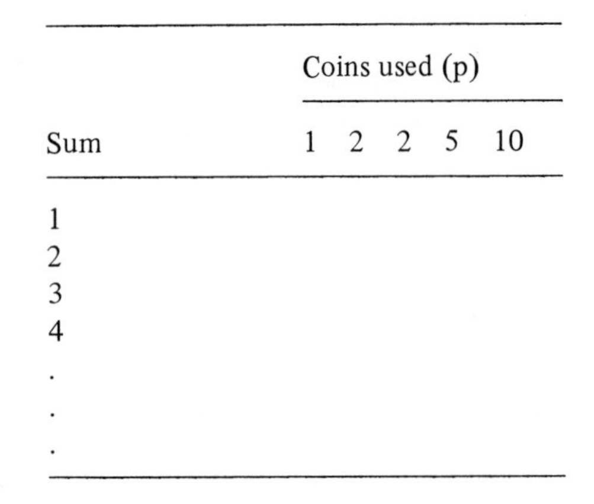

	Coins used (p)				
Sum	1	2	2	5	10
1					
2					
3					
4					
.					
.					
.					

Could you manage to make all the sums of money in whole pence from 1 p to 20 p using fewer coins?

B. Find the smallest number of coins needed to make all whole pence sums up to

(a) 10 p; (b) 15 p; (c) 45 p; (d) 50 p; (e) £1.

C. All the questions so far have been about writing numbers as terms added together, that is as *sums*. You can probably think of occasions when we can make a quantity by subtraction, that is using *differences*. Suppose you had an old-fashioned balance. You could weigh a 2-kg parcel using a 3-kg and a 1-kg weight (Fig. 8.5). In fact with weights of 1 kg and 3 kg, you could weigh four different quantities (Table 8.5). With a 1-kg and a 3-kg weight the heaviest object you could weigh would weigh 1 kg + 3 kg = 4 kg.

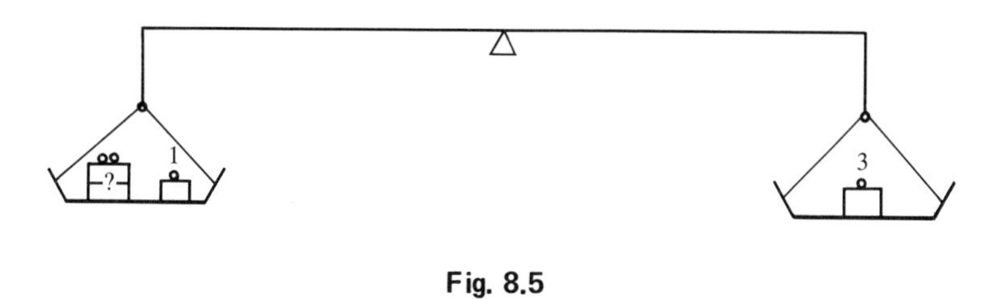

Fig. 8.5

Now suppose you had weights of 1, 3 and 9 kg. You should be able to weigh 1 + 3 + 9 different amounts. Make a table to show how it's done.

Table 8.5

Weight measured (kg)	Weights used in measuring	
	1-kg	3-kg
1	+	0
2	–	+
3	0	+
4	+	+

D. Could you have made thirteen different weights using *any* three, or is there something special about 1, 3 and 9? If you think you have found the pattern, try answering this old puzzle. A blacksmith has a bar of iron which weighs 40 pounds. How should he cut it so as to be able to weigh (on his old-fashioned balance!) any whole number of pounds up to 40?

8.6. Map colouring

Figure 8.6 shows a sketch map of Africa.

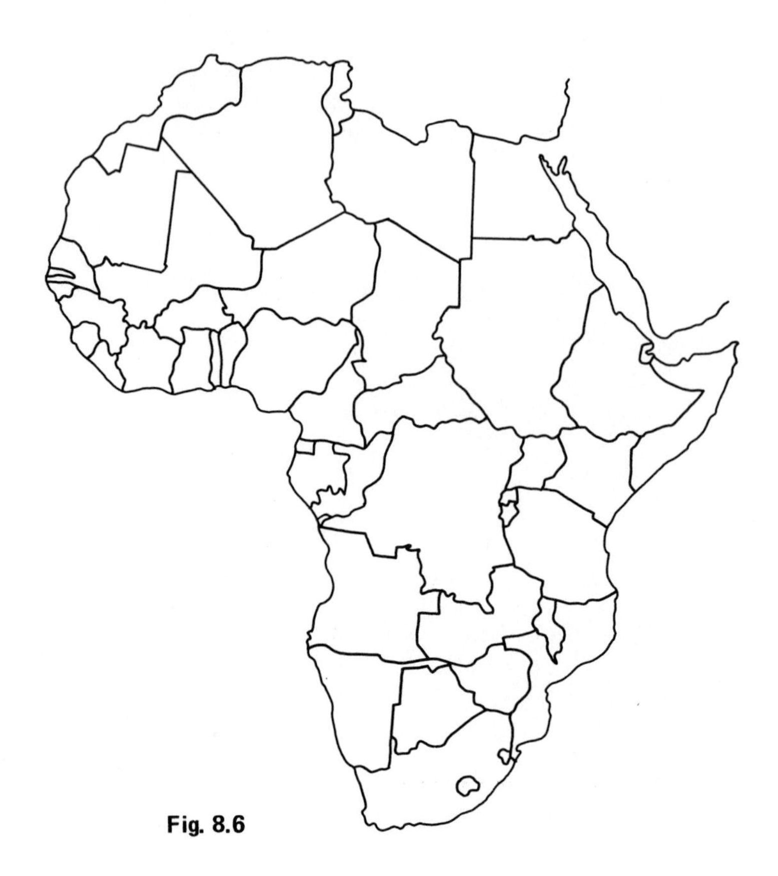

Fig. 8.6

What is the *smallest* number of colours which are needed in order to colour in this map so that countries with common borders do not use the same colour?

(Rather than actually colour in, mark the countries on the map with 1, 2, 3, . . . for different colours).

Fig. 8.7

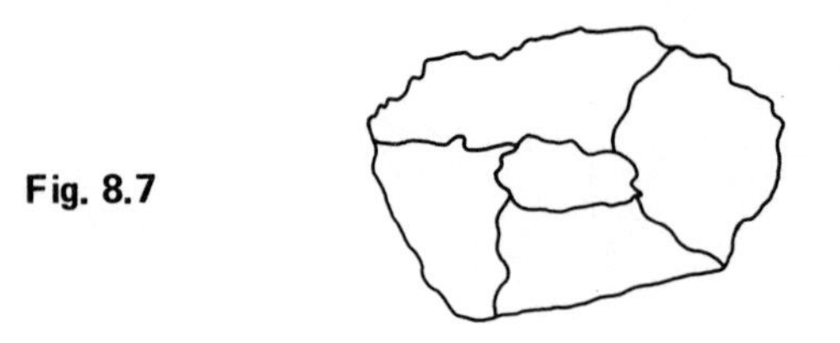

A simpler example of an island with five countries is shown in Fig. 8.7. If each country (and the sea) had a different colour, then we would need six colours. Can you use less? We actually used only four colours (Fig. 8.8).

Fig. 8.8

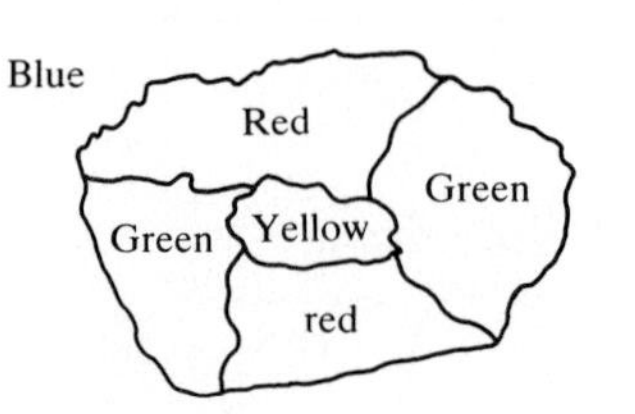

Now try the map of Africa (Fig. 8.6). Can you colour it using only four colours? If you cannot, try the following problem, again using only four colours if possible.

● Problem

Use as few colours as possible to colour in maps (a), (b), (c), (d), and (e) in Fig. 8.9.

Fig 8.9 here

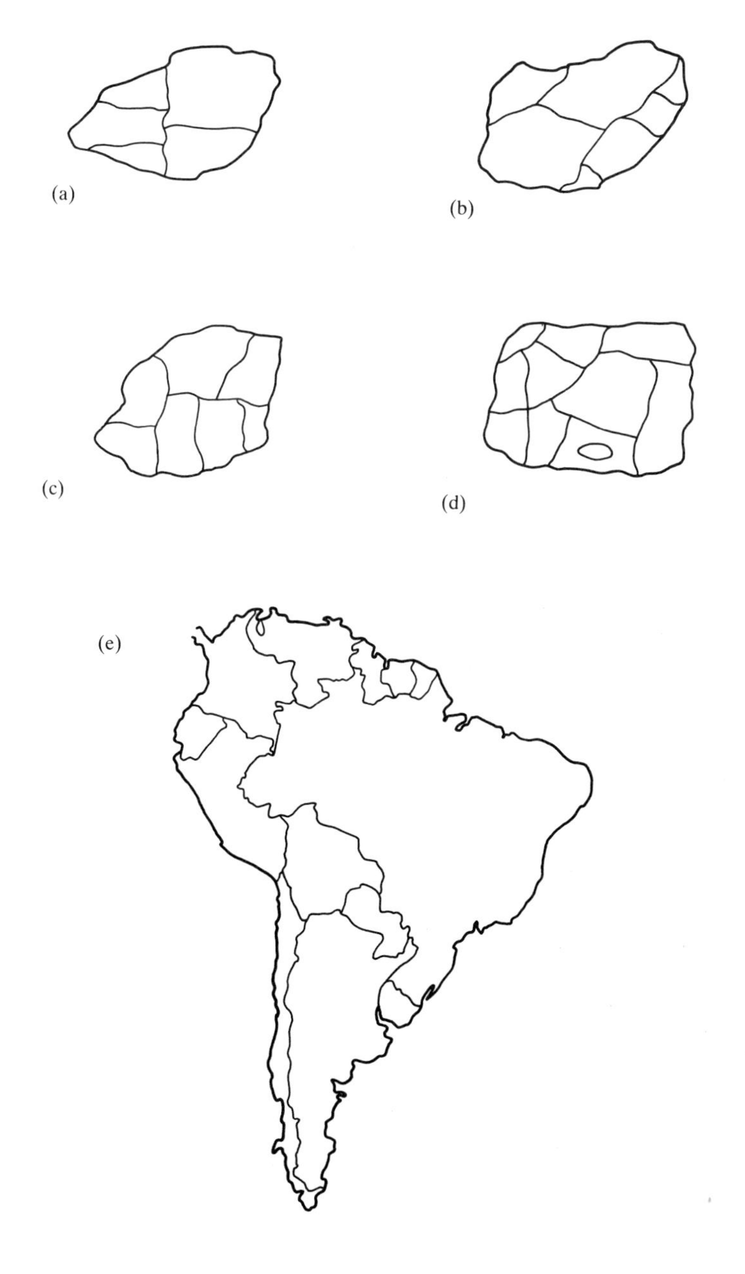

Fig. 8.9

Part II. Solutions to Problems

1 Sport

1.1. Super league football

1.

Aston Villa			Ipswich			Liverpool		
Win	Draw	Lose	Win	Draw	Lose	Win	Draw	Lose
3–1			3–2					1–3
		2–3			0–1		2–2	
2–0					2–4	4–2		
1–0					1–2	2–1		
	2–2		1–0			2–0		
	2–2		3–2					0–2
1–0			2–1			2–1		
3–1					2–3	1–0		
		1–2			2–5	3–1		
		0–3		1–1		1–0		
5 Won	2 Drawn	3 Lost	4 Won	1 Drawn	5 Lost	7 Won	1 Drawn	2 Lost
Points		17	Points		13	Points		22
Goals for		17	Goals for		17	Goals for		18
Goals against		14	Goals against		21	Goals against		12

Manchester City			Southampton			Tottenham		
Win	Draw	Lose	Win	Draw	Lose	Win	Draw	Lose
	2–2				0–1			0–2
3–0			2–1					1–3
		0–2			1–2	2–0		
		1–3			0–1			0–1
		0–1			0–1	3–2		
1–0				1–1		4–2		
		2–3			1–2	3–2		
		2–3	5–2				1–1	
	1–1			0–0			0–0	
		2–4			0–2	2–0		
2 Won	2 Drawn	6 Lost	2 Won	2 Drawn	6 Lost	5 Won	2 Drawn	3 Lost
Points		8	Points		8	Points		17
Goals for		14	Goals for		10	Goals for		16
Goals against		19	Goals against		13	Goals against		13

2.

Name of team	Games				Goals		
	Played	Won	Drawn	Lost	For	Against	Points
Liverpool	10	7	1	2	18	12	22
Aston Villa	10	5	2	3	18	14	17
Tottenham	10	5	2	3	16	13	17
Ipswich	10	4	1	5	17	21	13
Southampton	10	2	2	6	10	13	8
Man City	10	2	2	6	14	19	8

3.

Aston Villa	Ipswich	Liverpool
42 371	24 671	41 253
40 324	25 729	39 291
39 587	19 423	44 472
37 469	21 661	37 357
36 876	25 422	38 243
196 627	116 906	200 616

Man City	Southampton	Tottenham
34 421	21 012	46 290
47 281	26 775	47 203
39 420	28 031	39 241
35 624	28 833	43 870
27 801	30 674	39 783
184 547	136 325	352 712

● Related problems

A. £401 232
B. £2571 193.20
C. £881 780
D. £40 123.20
E. £1700
F. £1200
G. £2640
H. £700
I. £950
J. £780; £1300; second system.

1.2. Snooker

1. The maximum break arises by potting red followed by black 15 times, and then potting all the colours in the correct order. This gives a break of

$$\begin{aligned} 15 \quad & \times\ (1 + 7) + 2 + 3 + 4 + 5 + 6 + 7 \\ & = 15 \times 8 + 27 \\ & = 147. \end{aligned}$$

2. Player A's best break is to pot red, black, red, black, followed by all the colours in the correct order giving a break of

$$2\,(1 + 7) + 27 = 55.$$

Clearly A can still win the game.

3. Player A can still make a break of 27, giving a total score of 55, and so A can still win.

4. If the cueball hits the cushion only once, we have the possible paths shown in Fig. S1.1(a) which indicates three possible paths. If the cue ball can hit the cushion twice, we have the four possible paths shown in Fig. S1.1(b).

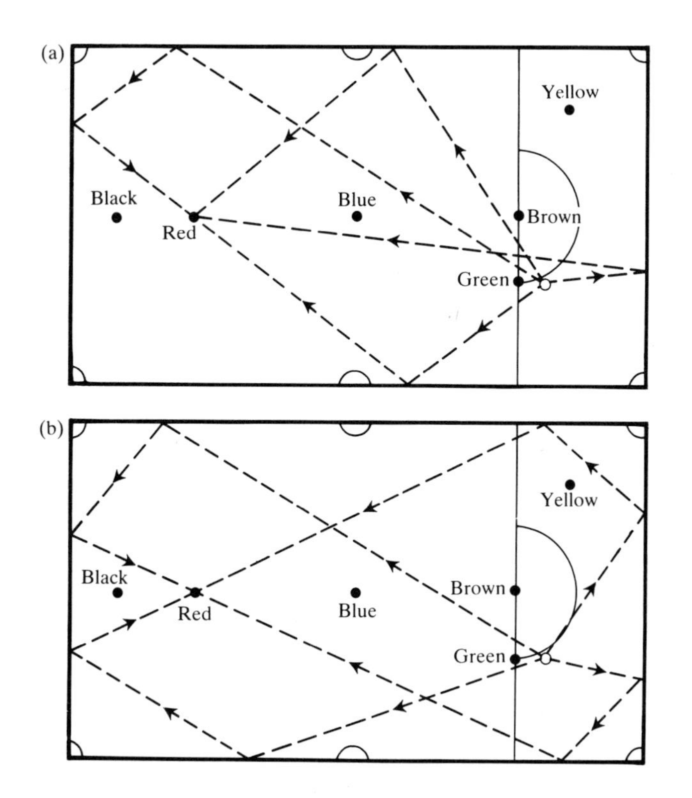

Fig. S1.1

1.3. Darts

1. Five ways of scoring 57 are

1,	6,	50
2,	5,	50
3,	4,	50
3 × 1,	4,	50
3 × 2,	1,	50.

In each case the first two throws could be in reverse order, giving 10 possibilities in all. Perhaps students should be encouraged not to be too concerned about the order of throws.

The total number of answers is very large. Students could be asked for say, 20 possibilities, or they might compete in pairs.

2. There are six ways of finishing. After 8 you might throw

9, 2 × 20 or 11, 2 × 19 or 13, 2 × 18 or 15, 2 × 17,
or 17, 2 × 16 or 19, 2 × 15.

3. Nine throws are needed (i.e. three visits to the ockey). Possibilities are

(3 × 19),	7 × (3 × 20),	(2 × 12)
(3 × 17),	7 × (3 × 20),	(2 × 15)
(3 × 15),	7 × (3 × 20),	50, (2 × 17)
(3 × 17),	6 × (3 × 20),	50, (2 × 20).

These are not asked for, but interested students could be asked to find them.

4. Eleven throws are needed.

5. The best strategy would be to aim for treble 14, when a near miss still gives a good score. The sort of calculation which might be given is

Aiming at 20		*Aiming at 14*	
Three throws	(20	Three throws	(14
	(1		(11
	(5		(12
	26		37

These totals might be given as average scores for three throws in the two cases, (ignoring trebles, which are unnecessary complications). But it would be a more accurate interpretation of the situation described if the calculations were

Average for four throws

Aiming at 20	*Aiming at 14*
20	14
1	11
20	14
5	12
46	51

14 still has the edge on 20, resulting in an average per throw of 12¾, against 11½, not counting trebles.

The only other strategy likely to be proposed is to throw for the bull. Any criticism would lead to a comparison of areas. If the bull and outer are missed entirely – which is more probable then missing a treble – the average score is 10½ and, of course, there is no hope of a treble. Perhaps a practical test is the only way of deciding the matter!

● Related problems

Students might be encouraged to group their matches, for example,

4 players (A, B, C, D)

A x B		
	B x C	
A x C		C x D
	B x D	
A x D		

5 players (A, B, C, D, E)

A x B			
	B x C		
A x C		C x D	
	B x D		D x E
A x D		C x E	
	B x E		
A x E			

and should eventually realize that with n players, the number of matches is

$1 + 2 + 3 + \ldots + (n - 1)$
12 players – 66 matches
20 players – 190 matches.

The more elusive formula $\frac{1}{2}n\,(n - 1)$ might be pointed out in the case of some students.

1.4. Arranging a knockout tournament

1. The competition might go like this

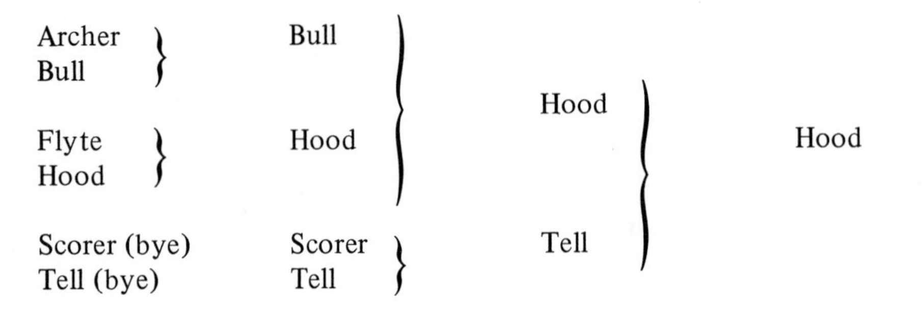

Again there are 5 games in 3 rounds.

2. 3 rounds, 7 games.

3. 4 rounds, 8 games. Note the byes can be anywhere.

4. Students may need extra examples before they find the rule for the number of rounds, but should easily spot

$$n \text{ players} - (n-1) \text{ games}$$

and see the reason why. The number of rounds will probably be found by continual halving, counting the next whole number upward where necessary, e.g.

$$9 \to 5 \to 3 \to 2 \to 1 \text{ (4 rounds)}$$
$$14 \to 7 \to 4 \to 2 \to 1 \text{ (4 rounds)}$$

The sophisticated formula

$$[\log_2 (n-1)] + 1 \text{ (Where square brackets represent 'the integral part')}$$

need not be mentioned.

Answers to examples in 4.

(a) 4, 13; (b) 5, 20; (c) 6, 32.

● Related problems

(a) 66; (b) 36; (c) 11.

A knockout tournament always produces an outright winner but runners-up are not ranked. In a Swiss tournament a good ranking is produced but competitors may tie with the same score. To resolve this, either extra games must be played or a rule introduced whereby if two players have the same final score the result in any game in which they met determines which of them has precedence.

1.5. Cribbage

Pupils would find it helpful to use actual packs of cards.

1. (i) 7 points; (ii) 10 points; (iii) 11 points; (iv) 5 points;
(v) 8 points; (vi) 4 points; (viii) 4 points; (viii) 4 points;
(ix) 10 points; (x) 0 points.

3. (i) jack hearts (4 points); (ii) 4 spades (12 points); (iii) jack diamonds (4 points); (iv) choice between queen diamonds and 4 diamonds (6 points) (keeping the 4 diamonds is better because the hand has greater potential with a turn-up card); (v) 9 diamonds (6 points); (vi) choice between 4 spades and 9 hearts (4 points) (keeping the 4 spades is slightly better); (vii) choice between jack spades and ace spades (5 points) (keeping the ace spades is slightly better); (viii) choice between 4 hearts and 7 clubs (6 points) (keeping the 7 clubs is better); (ix) 6 hearts; (x) 2 clubs.

● Related problems

A. 8 points; 2 . 8 points; 3. 7 points;
B. 16 points; 5. 12 points.
C. Totals 2, 0, 3, 5, 2.
D. 5 diamonds.

1.6. Matchstick men

A. The initial number of matches is unimportant, since you can always win as long as it's your opponent's turn when 4 matches remain

Him 1, You 3
Him 2, You 2
Him 3, You 1

Thus your strategy is the same; leave a multiple of 4 matches.

B. In this case you still want to be able to win whatever your opponent takes.

Him 2, You 4
Him 3, You 3
Him 4, You 2

Thus your strategy is to leave a multiple of 6 matches all the time.

C. The reasoning above shows that the number 35 is irrelevant. The winning plays are

Him 4, You 6
Him 5, You 5
Him 6, You 4

and so the best strategy is to leave multiples of 10 each time.

D. As above, leave 11 each time. Notice that

$$4 = 1 + 3$$
$$6 = 2 + 4$$
$$10 = 4 + 6$$
$$11 = 4 + 7$$

and so the minimum and maximum choice seems the key.

E. Size of initial heap unimportant.
Strategy is to leave (minimum or maximum) matches each time.

1.7. Bets with a die

1. Yes.
2. No.

● Related problems

A. $2 \times £5 - 4 \times £2 = £2$. Yes.

B. $1 \times £12 - 5 \times £3 = -£3$. No.

C. $4 \times £4 - 2 \times £8 = 0$. Just no!

2 Travel

2.1. A holiday abroad

1. Hotel Betiana

2 adults at £226	£452.00
1 child at adult cost of £226	£226.00
1 child at £158 less 10 per cent	
£158 – £15.80	£142.20
Airport charges for 4 people	
£9.20 X 4	£ 36.80
Total	£857.00

Hotel Galien

2 adults at £237	£474.00
1 child at adult cost of £237	£237.00
1 child at £145 less 10 per cent	
£145 – £14.50	£130.50
Airport charges for 4 people	
£9.20 X 4	£ 36.80
Total	£878.30

Hotel Alphina is the least expensive.

Cost of 7-day holiday between 1st May and 13th May:

2 adults at £193		£386.00
1 child at adult cost of £193		£193.00
1 child at £125 less 10 per cent		
£125 – £12.50		£112.50
Airport charges for 4 people		
£9.20 X 4		£ 36.80
Total		£728.30
Saving of £835.40 – £728.30	=	£107.10.

2. Cost of 14-day holiday between 1st May and 13th May.

2 adults at £249	£498.00
1 child at £249	£249.00
1 child at £137 less 10 per cent	
£137 – £13.70	£123.70
Airport charges for 4 people	
£9.20 X 4	£ 36.80
	£907.50

Cost of 14-day holiday – cost of 7-day holiday = £907.50 – £728.30 = £179.20.

● Related problems

A.

BIRMINGHAM–LUTON

For a one-week holiday all flights are from Luton.

Return Birmingham–Luton bus fare is £5.50 for adults and £3.85 for children (under 16).

Thus, the cost for the Jones family is

$$£2 \times 5.50 + £2 \times 3.85 = £11 + £7.70 = £18.70$$

COST OF PACKAGE HOLIDAY

Cost for 1 adult = £155.

Children receive reductions which depend on age of child, category of hotel, and when the holiday is taken. The younger child gets a 50 per cent reduction. Therefore,

Cost is 50 per cent of £155
$$= 50/100 \times £155$$
$$= \tfrac{1}{2} \times £155$$
$$= £77.50.$$

The older child gets a 10 per cent reduction. Therefore,

Cost is 90 per cent of £155
$$= 90/100 \times £155$$
$$= 9/10 \times £155$$
$$= £139.50.$$

The total cost of the package is

$$£155 + £155 + £77.50 + £139.50 = £527.$$

AIRPORT CHARGES

A charge of £9.45 is added to cover airport taxes.

INSURANCE

Insurance is recommended and costs £5.95 per person. The total insurance cost is £23.80. Thus, the grand total is

$$£18.70 + £527.00 + £9.45 + £23.80 = £578.95.$$

B. Bus fare is the same.

Cost of package

Cost for 1 adult = £150. The younger child goes free. The older child received a 10 per cent reduction. Therefore, the cost is 90 per cent of £150 = £135.

Therefore, the total cost of the package is

£150 + £150 + £135 = £435.

The aiport and insurance charges are the same.

The grand total is

£18.70 + £435.00 + £9.45 + £23.80 = £486.95.

C. For a 2-week holiday, the family can either fly from Birmingham or catch the bus to Luton and fly from there. For convenience they decide to fly from Birmingham.

Cost of package for 1 adult = £239.

The younger child receives a 50 per cent reduction. Therefore, the cost for the younger child is

$50/100 \times$ £239 = £119.50

The older child receives a 10 per cent reduction. Therefore, the cost for the older child is

$90/100 \times$ £239 = £215.10

Total cost of package is £239 + £239 + £119.50 + £215.10 = £812.60.
Insurance charges are £7.30 per person, i.e. £29.20 in total.
Airport charges are £9.45.

The grand total is, therefore,

£812.60 + £29.20 + £9.45 = £851.25.

2.3. British Rail timetables

Examples and times given refer to 1985/86 British Rail timetable.

1. 12.59.
2. 13.40.
3. Tables required are 26 and 39. Using the 1985/86 timetable we have: first train after 11.00 is 11.11. Arrival time in York 11.44. Leave York at 11.51. Arrive Leeds 12.18.

2.5. Youth hostelling

1.

	Young	*Junior*
Matlock Bath	£1.80	£2.15
Windgather Cottage	£1.30	£1.65
Hagg Farm	£1.80	£2.15
Bakewell	£1.80	£2.15
	£6.75	£8.10

2.

	Young	*Junior*	*Bag Hire*
Osmotherley	£2.05	£ 2.50	65 p
Helmsley	£1.80	£ 2.15	65 p
Lockton	£1.30	£ 1.65	60 p
Boggle Hole	£2.05	£ 2.50	65 p
Whitby	£1.80	£ 2.15	65 p
	£9.00	£10.95	£3.20

Total cost for young member = £9.00 + £3.20 = £12.20.

Total cost for junior member = £10.95 + £3.20 = £14.15.

3 Parties

3.1. How much from a bottle?

1. 13.
2. 21.
3. 7.
4. 12.
5. 5 pints of beer; 10 pints lemonade.
6. 6; 315 fl oz.
7. Better.
8. Not cheaper than first, but cheaper than second.

3.2. Can you cook?

1. 55 p if 4 portions; 44 p if 5 portions; 36 p if 6 portions.
2. Approximately 80 p.

● Related problems

A. 200 g butter; ½ a lemon; 950 g caster sugar; 75 g chocolate.

B. £4.99.

C. No the conversions are not the same. Ingredients are usually given in multiples of 2 oz or 25 g. It is not usually a good idea to try and convert recipes exactly.

D. Time, presentation, storage, taste, quality, individual preference.

3.3. Cost of food

1. (a) £5.53; (b) £6.96; (c) £13.19; (d) £7.85.
2. £1.20 + £1.35 + £4.32 + £5.40 + £1.56 + £2.50 = £16.33.
3. £4.64 + £8.64 + £0.78 = £14.06. No.
4. £2.76 + £4.40 + £1.35 = £8.51. Two boxes.
5. £0.40 + £1.36 + £1.10 + £0.39 + £0.65 = £3.90. 6 hours.
6. (a) 3.9 p; 4.3 p; first; (b) 64½ p per 100 g; 66½ p per 100 g; first.
7. 44 p per 100 (type A); 72 p per 100 (type B); 49.5 p per 100 (type C). Type A.
8. Yes. The last.
9. The third; the second.

3.4. Running a bar at a disco

1. Profit for wine £28 – £17.50 = £10.50. Profit for beer = £27 – £9 = £18. Total profit = £37.50.

2. *Disco A*

	Takings (£)	Cost (£)	Profit (£)
Beer	36.00	33.00	3.00
Wine	25.00	24.50	0.50
Coke	20.00	10.20	9.80
Total			13.30

Disco C

	Takings (£)	Cost (£)	Profit (£)
Beer	32.40	21.00	11.40
Wine	22.50	14.00	8.50
Coke	30.00	7.80	22.20
Total			42.10

The price list for Disco C gives the greatest profit; the prices are not yet so high that sales have been too seriously reduced. Coke in particular is holding up well, probably because it is the cheapest drink available.

A. 29 p X 135 glasses = £39.15 which gives the greatest takings.

B. *Wine*

25 p X 100 glasses = £25
34 p X 82 glasses = £27.88
35 p X 80 glasses = £28 (the greatest takings)
45 p X 50 glasses = £22.50

Coke

10 p X 200 = £20
15 p X 100 = £27
20 p X 150 = £30

There is no evidence that a price greater than 20 p will further increase takings. Extrapolating the graph *suggests* 21 p X 142 glasses = £29.82, i.e. slightly *worse* than 20 p.

3.5. Don't drink and drive

1. 10.
2. 6.
3. 60 mg.
4. 8 hours.
5. No.
6. (a) 2½ pints; 1½ pints; none.
   (b) 5 glasses, 3 glasses, none.
7. See Fig. S3.1(a).
8. See Fig. S3.1(b).

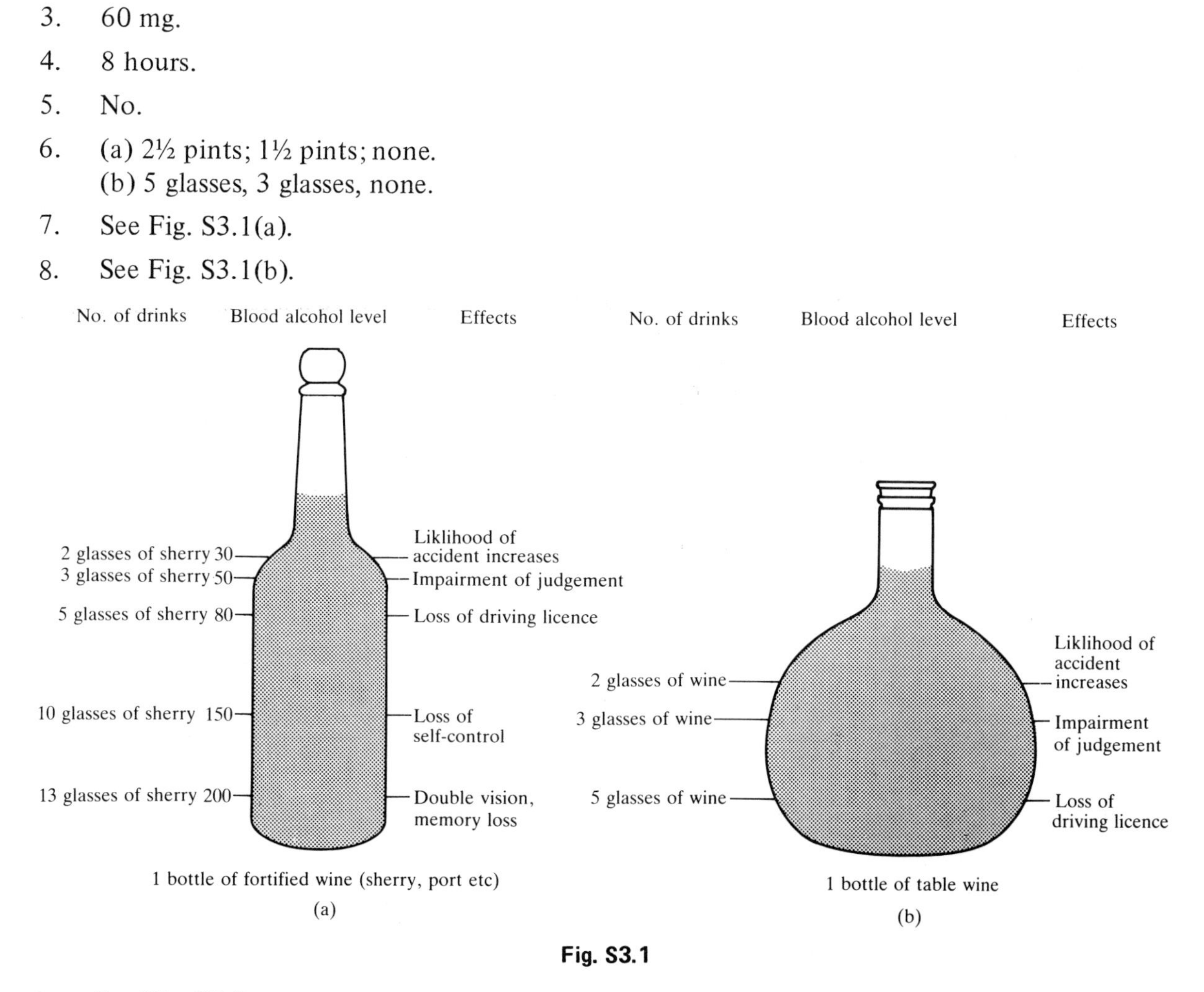

Fig. S3.1

9. See Fig. S3.2.

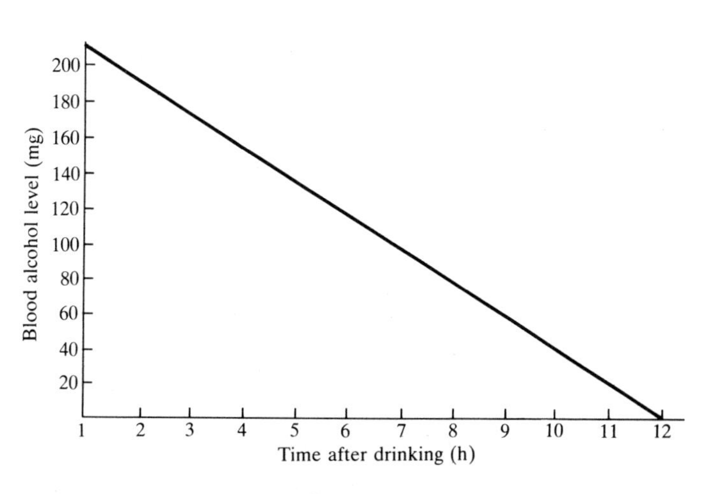

Fig. S3.2

10. Any work with machinery.

4 Money

4.1. Premium bonds

1. *Each week.*

1 @	£100 000.00
1 @	£ 50 000.00
1 @	£ 25 000.00
	£175 000.00

Each month.

1 @	£250 000	=	£ 250 000.00
5 @	£ 10 000	=	£ 50 000.00
50 @	£ 5 000	=	£ 250 000.00
250 @	£ 1 000	=	£ 250 000.00
750 @	£ 500	=	£ 375 000.00
25 000 @	£ 100	=	£2 500 000.00
75 000 @	£ 50	=	£3 750 000.00
			£7 425 000.00

Each year.

£ 175 000 × 52	=	9 100 000.00
£7 425 000 × 12	=	89 100 000.00
		98 200 000.00

2. Total prize money distributed each year is £98 200 000 which represents 7 per cent of total investment. Therefore,

$$\text{Total investment} = \frac{98\ 200\ 000}{7} \times 100 = £1\ 402\ 850\ 000.$$

3. No. of prizes offered each year = 52 × 3 weekly prizes + 12 × 101 056 monthly prizes = 156 + 1 212 672 = 1 212 828 prizes.

4.2. Time cards

Hours worked by Joe Smith

Monday	8.35	
Tuesday	8.54	
Wednesday	9.15	Total 45.00
Thursday	9.17	
Friday	8.59	
Saturday	–	
Sunday	3.30	

Thus we have the table

	Basic	Overtime	Sunday	
Hours/mins	35.00	10.00	3.30	
Rate £	2.00	3.00	4.00	Gross pay
Pay £	70.00	30.00	14.00	114.00

● Related problems

A. Hours worked

Monday	8.48	
Tuesday	8.23	
Wednesday	9.11	Total 43.03
Thursday	8.00	
Friday	8.41	
Saturday	–	
Sunday	2.44	

giving

	Basic	Overtime	Sunday	
Hours/mins	35.00	8.03	2.44	
Rate £	2.00	3.00	4.00	Gross pay
Pay £	70.00	24.15	10.93	105.08

B.

	Basic	Overtime	Sunday	
Hours/mins	35.00	10.00	3.30	
Rate £	2.20	3.30	4.40	Gross pay
Pay £	77.00	33.00	14.52	124.52

	Basic	Overtime	Sunday	
Hours/mins	35.00	8.03	2.44	
Rate £	2.20	3.30	4.40	Gross pay
Pay £	77.00	26.56	12.03	115.59

C.

	Basic	Overtime	
Hours/mins	36.00	4.23	
Rate £	1.72	2.12	Gross pay
Pay £	61.92	9.29	71.21

Please note that, for example, in C above 4.23 hours/mins means 4 hours 23 minutes and to calculate the second column, we must convert it to a decimal number; i.e.

4 hours 23 minutes = 4 + 23/60 hours = 4.38 hours

and we can now calculate the pay from

$4.38 \times 2.12 = 9.32$ (to two decimal places).

4.3. Savings

1. (i) After a year you have £113.93 p.
   (ii) £114.29 p.
   (iii) The interest after tax is £8.75 p. If income tax had not been deducted the interest would be £8.75 × 100/70 = £12.50 p.
   (iv) £15.38 p.

2. (i) $£100 \times (1.085)^5 = £150.37$ p
   (ii) $£100 \times (1.085)^{10} = £226.10$ p
   (iii) The compound interest at 8.75% is £52.11 p.
   The extra interest is £1.74 p
   (iv) The compound interest at 9.75% is £59.23 p
   The extra interest is £8.86 p.

3. (i) $£100 \times (1.0425)^4 = £118.11$ p
   (ii) $£300 \times (1.0425)^{10} = £454.86$ p
   (iii) £300 for 5 years becomes £460.35 p
   £500 for 10 years becomes £1177.33 p
   £1000 for 15 years becomes £3613.22 p

● Related problems

A. The first £70 of interest is free of all UK income tax (including investment income surcharge) and capital gains tax. This means that up to £1400 can be held tax-free. Husbands and wives are separately entitled to a tax concession.

B. 6.3 per cent and 7 per cent.

C. 'Interest will be charged at the rate of 2.25 per cent per month on a daily basis equivalent to a maximum annual rate of 30.6 per cent.'

4.4. Cash help

1. £21.15.
2. £47.35.
3. £26.45.
4. £47.05.
5. No.
6. £43.05.
7. £7.90.
8. £12.65.
9. £555.60.
10. Each would lose £2.55 benefit but their rent and rates would be paid.

5 Home

5.1. Independence is expensive

1. There are several types of accommodation which have different features. Here are some examples.

Digs	'Basic' furnished bedroom (sometimes shared) Share bathroom and toilet facilities No proper kitchen Meals provided
Hostels	Sometimes flats, sometimes single rooms Share bathrooms and toilet facilities Share bedrooms sometimes No kitchens Meals provided
Bedsit	One room –furnished Share bathroom, toilet, and kitchen
Flat	Several rooms usually unfurnished Often shared with a group of others.

The rent will depend on where you live, however the following list gives some ranges which applied in 1982.

Digs	£35–£45 London £15–£20 Provinces
Hostels	£15–£30 London £50–£60 YMCA/YWCA
Bedsit	£25–£30 upwards (London) £15 upwards (Provinces) (plus heating possibly)
Flat	£10–£20 depending on single or shared plus gas, electricity, etc. (Provinces) A lot more (London.)

2. Costs: New/old furniture for (a) kitchen; (b) living area; (c) bedroom.
3. Costs of heating, cooking, eating, lighting, living?

5.2. Furniture moving the easy way

TEACHER'S NOTES

If pupils have difficulty in making scale drawings then the exercise in the appendix should be attempted first. These questions are based on the two examples given in the text. Problems 1–4 require 2 sheets of A4 graph paper.

● Exercises

Lengths drawn should be:

1. (i) 5 cm; (ii) 10 cm; (iii) 4.5 cm; (iv) 3.8 cm; (v) 7.2 cm.
2. (i) 2 cm; (ii) 3 cm; (iii) 5 cm; (iv) 9 cm; (v) 2.5 cm.

Squares should have sides of length

3. (i) 2 cm; (ii) 4 cm; (iii) 6 cm; (iv) 5 cm; (v) 2½ cm.

5.3. Ready pasted?

1. Distance round room = 2 × (4 + 3) = 14 m; number of rolls required = 7; cost of wallpaper = £24.50.
2. Distance round room = 2 × (3.4 + 3) = 12.8 m – round up to 13 m; number of rolls required = 6; cost of wallpaper = £17.10.
3. Distance round room = 2 × (4.2 + 3.5) = 15.4 m – round up to 16 m or down to 15 m (makes no difference)
   Number of rolls required = 7.
   Cost of wallpaper = £8.05.
   Cost of paint = 2 × £2.70 = £5.40.
   Total cost of decorating = £8.05 + £5.40 = £13.45.
4. (a) Distance round outside of room = (8 + 5 + 3.5 +2 + 5.5 + 3) m = 26 m.
   Number of rolls cannot be read from table.
   Number of rolls required for half the room = 7; number of rollls for whole room = 2 × 7 = 14.
   Easily enough, possibly an overestimate.
   Cost of wallpaper = £74.20.

   (b) Number of rolls for 8 m wall = 4; cost = £21.20.
   Number of rolls of woodchip paper = 9; cost = £10.35.
   Cost of paint = £6.75.
   Total cost = £21.20 + £10.35 + £6.75 = £38.30.
   Amount saved = £74.20 – £38.30 = £35.90.

● Related problems

A. Total number of rolls is 33 or 34 depending if the lounge uses woodchip or not. The same strength paste is used for normal and woodchip papers. The packet will provide paste for 11–13 rolls so that 3 packets will be needed.

B. Area of door = 200 cm × 83 cm = 16 600 cm^2
Area of window = 135 × 80 cm = 10 800 cm^2
Total area of paper wasted = 27 400 cm^2
Length of paper wasted = 27 400/50 cm^2 = 548 cm
= 5.48 m. i.e. ½ roll wallpaper.

D. Teaching suggestion. This is an open question that can be answered simply (i.e. buy an extra roll) or using varying degrees of calculation. One way of looking at it is to imagine that the room is taller than it is, i.e. make the room have a height that is a whole multiple of pattern drops.

Thus, if the room is in fact 2.2 m and the pattern drop is 30 cm, then we consider the room as being 2.4 m (i.e. 8 × 30 cm) and use Table 5.2 as before.

● Exercises

1.

Height from skirting (m)	Measurements of walls (m)	Number of rolls
2.2	12	6
2.6	18	10
2.0	9	4
3.0	14	9
2.8	19	11
2.4	10	5
2.6	11	6
3.0	18	11

2. (a) 12 m; (b) 14 m; (c) 16 m.
3. (a) 2.2 m; (b) 2.6 m.

5.4. Tape it

1. The total length of tracks to be taped is
   18 + 20 + 2 + 2½ + 3 + 3½ + 4 + 4½ + 5
   = 62½ minutes.

So it is *not* possible to record everything on one C60 cassette.

2. 18 + 3 + 4 + 5 = 30;
   20 + 2 + 3½ + 4½ = 30.
3. 20 + 18 + 3 + 4 = 45;
   19½ + 21 + 2 + 2½ = 45 Yes.
4. 3 C60 tapes cost £3.45. 2 C90 tapes cost £3.10.
5. Dylan side 1 + Dylan side 2 = 42:42 minutes.
   Joseph side 2 + Streisand side 1 = 43:42 minutes.
   Streisand side 2 + Joseph side 1 = 43:22 minutes.
   Harvest side 1 + Harvest side 2 + 2 singles = 44.50 minutes.
   } 2 C90 tapes.

6 House design

6.2. Describing houses

1. Estate agents include:–

facilities in the town	size and standard of garden	rateable value
local facilities	standard of maintenance	location
size	standard of decoration	freehold/leasehold
type	description of rooms	special facilities
age	heating	

2. The local council would probably limit itself to:–

location	size	no of bedrooms	rent
which floor (if	heating rates (if not included in rent)		
flat or maisonette)	telephone	size of garden	

6.3. How well do you know your house?

Look out for
- long thin rooms, especially airing cupboards
- unusable stairs
- gaps between rooms

6.4. Building costs

The solutions to these sections depend on the students' designs.

6.6. What next?

		Prerequisites	Estimated time seconds
A	Get bread	–	20
B	Cook toast	A	120
C	Butter toast	B, D	40
D	Get plate and knife	–	20
E	Heat beans	–	60
F	Put beans on toast	C, E	10

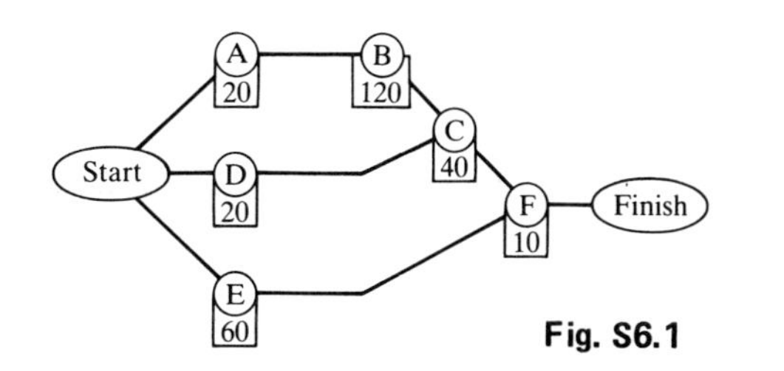

Fig. S6.1

The network for making beans on toast is shown in Fig. S6.1. Critical path = A, B, C, F. Total time = 190 seconds (estimated).

● Related problems

Building a house

	Operation	Done after	Time taken (days)
A	Clear site	–	1
B	Dig and lay drains	A	3
C	Dig and concrete four oversite	B	3
D	Build walls up to f	C	5
E	Put in downstairs frames	D	½
F	Put in floor joists	D	2
G	Build walls up to roof	F	5
H	Put in upstairs window and door frames	G	½
I	Fix roof timbers and felt	G	3
J	Tile roof	I	3
K	Lay floors and build staircase	Q. I, N	4
L	Hang doors and fit handles etc	I	1
M	Install electrics	I	3
N	Install heating	G	4
O	Glaze windows	E, H	1
P	Plaster	M, O	5
Q	Fit plumbing	G	8
R	Fit skirting boards and floor architraves	L, P	3
S	Paint outside	G	4
T	Fix gutters and downspouts	J	1
U	Fit kitchen cupboards etc	R	2
V	Tile kitchen and bathroom	K, U	2
W	Paint inside	V, P + 3 weeks!	5

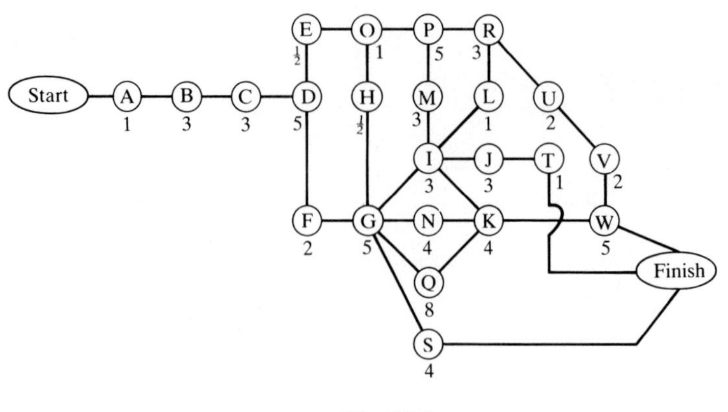

Fig. S6.2

The network is shown in Fig. S6.2. Critical path is A, B, C, D, F, G, Q, K, W. Time = 36 days. Usually takes much longer because too many operations cannot go on at once in a limited space.

7 Buying

7.1. How much off?

The method of solving the problem depends on the ability of the pupil.

1. £3.91.
2. £165.60.
3. Hill and Son (£4.20 compared with £4.25).
4. £320.
5. The Car Insurance Co (£175 compared with £180).
6. £207.
7. £532.
8. £726.
9. Monthly £9.60, half-yearly £8.40, yearly £8.16.

EXTRA WORK

A. 1. £35; 2. £467.50; 3. £2.21; 4. £13.94; 5; £204; 6. £328.10; 7. £389.30; 8. £3.15; 9. £39.20; 10. £245; 11. £399; 12. £111.30; 13. £38.40; 14. £42; 15. £105; 16. £216.

B. 1. £17; 2. £25.50; 3. £42.50; 4. £76.50; 5. £68; 6. £50.50; 7. £12.75; 8. £55.25; 9. £80.75; 10. £72.25; 11. (a) £16; (b) £28; (c) £44; (d) £56; (e) £76.

7.2. Discount groceries

1. Savings = 16 p + 8 p + 14 p + 24 p = 62 p.
3. Spar total £13.69; Asda total £12.61; saving £1.08.

Place	Distance to Yate	Distance there and back	Cost (£) (at £0.04 per mile)	Worth the journey	Cost (£) (at £0.038 per mile)
Alveston	8	16	0.64	Yes	0.61
Chepstow	13	26	1.04	Yes	0.99
Ch. Sodbury	2	4	0.16	Yes	0.15
Avonmouth	14	28	1.12	No	1.06
Portishead	19	38	1.52	No	1.44
Keynsham	12	24	0.96	Yes	0.91
Bath	16	32	1.28	No	1.22
Chippenham	17	34	1.36	No	1.29

4. £1.52/40 gives 3.8 p per mile. Avonmouth is now just worth the journey.

7.3. A good deal?

The method of solving the problem depends on the ability of the pupil.

1. Music for You Ltd (£46 compared with £47).
2. Hill's Washing Machine Co (W.J. Mansell & Son's quote is £103.50).
3. F. Wall & Son (Brian J. Painter Ltd's quote is £862.50).
4. £264.50.
5. Single-glazing costs £310.50.
6. New Drives Ltd (Roads and Bridges can be ignored, Path & Drives Ltd quote is £632.50).
7. J.W. Electrics Ltd (£747.50, F. Hill & Sons can be ignored).
8. £672.
9. £412.50.
10. £141 600.
11. 51 p.

● Exercises

A. 1. £138; 2. £402.50; 3. £46; 4. £161; 5. £207; 6. £299;
7. £506; 8. £34.50; 9. £103.50; 10. £172.50; 11. £310.50; 12. £356.50.

B. 1. £69; 2. £138; 3. £207; 4. £299; 5. £437; 6. £529;
7. £103.50; 8. £241.50; 9. £333.50; 10. £379.50; 11. £563.50.

7.4. Buying by instalments

1. £74.
2. £18.
3. £47.40.
4. £97.80.
5. £37.56.
6. You pay £63. Your friend pays £63. Neither: both pay the same.

7.5. The cost of advertising

1. (a) 96 p; (b) £1.28; (c) 96 p; (d) £1.28; (e) 96 p; (f) £1.28; (g) 96 p; (h) £1.28; (i) £1.28.
2. (a) £2; (b) £4; (c) £20.
3. (a) £12; (b) £48; (c) £20.
4.

Advert	No. of columns	No. of cm long	No. of column-cm	Cost (£) at £2 per column-cm
(a)	1	8	8	16
(b)	1	6	6	12
(c)	2	5	10	20
(d)	1	5	5	10
(e)	3	6	18	36
(f)	2	4	8	16

A. Block advertisements

1. (a) £5000; (b) £2500; (c) £1250; (d) £625; (e) £2500.
2. (a) 160; (b) £62.50.

(c)

A. £125	G. £437.50	M. £750	S. £875
B. £250	H. £312.50	N. £125	T. £125
C. £187.50	I. £625	O. £62.50	U. £125
D. £750	J. £125	P. £125	V. £750
E. £250	K. £937.50	Q. £375	
F. £562.50	L. £250	R. £1312.50	

B. TV advertising

1.

Monday	12 hours
Tuesday	13 hours
Wednesday	13 hours
Thursday	12½ hours
Friday	14 hours
Saturday	14½ hours
Sunday	13 hours
Total	92 hours

2. 828 minutes.
3. £6 624 000.
4. £344 450 000.
5. £387 500 000.
6. (a) £7 382 250; (b) £383 880 000.

8 Odds and ends

8.1. 20 questions

1. Most likely to be 60–74 ins (B)
2. Most likely to be 150–190 cm (C)
3. Most likely to be 126–146 lb (C)
4. Most likely to be 60–70 kg (C)
5. 2.8 cm (C)
6. Most likely (B)
7. A bag of crisps weighs about 25 g so (B)
8. 25 litres (A)
9. 575 ml (check) (A)
10. 28 g (D)
11. £73 (C)
12. Probably 0.05–0.1 mm (A)
13. 25 hours/wk for 40 weeks; i.e. 1000 h (D)
14. 17 years = 6205 days (B)
15. 1985 years = 724 525 days (D)
16. Depends on the room
17. 18 m/s (C)
18. 400 000 km (C)
19. 16.5 km/1 (B)
20. 1 300 000 approximately (C)

8.2. Birthdays

1. 4 June 1972 Y = 1972.
   D = 31 + 29 + 31 + 30 + 31 + 4 = 156
   $\frac{Y-1}{4} = \frac{1971}{4}$ = 492 (remainder 5).
   S = Y + D + quotient $\frac{(Y-1)}{4}$ − 15
   = 1972 + 156 + 492 − 15
   = 2520 − 15
   = 2505.
   2505/7 = 357 (remainder 6 – Friday).

2. 29 July 1981 Y = 1981.
   D = 31 + 28 + 31 + 30 + 31 + 30 + 29 = 210.
   $\frac{Y-1}{4} = \frac{1980}{4}$ = 495.
   S = 1981 + 210 + 495 − 15
   = 2686 − 15
   = 2671.
   2671/7 = 381 (remainder 4 – Wednesday).

3. Monday.

5. Wednesday.

6. Sunday.

7. Sunday.

8.3. Queues

TEACHING HINTS

A simple way to investigate this type of problem is by simulation. Both the given problems can be simulated by using a die. Some discussion might be useful to see how to represent probabilities of 1/6, 1/3, and ½.

More able pupils may be able to understand the analytical argument.

ANALYTICAL SOLUTION

On average ½ person arrives every minute. That is, one person arrives every two minutes. Once there is someone in the queue we have, on average, ½ a person joining and 1/3 of a person leaving every minute. This is a net gain of 1/6 person per minute. Thus after the first 2 minutes the queue will increase in size at the rate of 1/6 person per minute. Hence, after a total of 20 minutes, we have 1 person arriving within the first two minutes and 1/6 of a person arriving per minute for the next 28 minutes. So the length of the queue is

$$1 + 1/6 \times 18 = 4 \text{ people.}$$

● Problem solutions

1. 20 minutes; a further 6 minutes.
2. 7.8 people; 22 minutes.
3. 6.8 people.
4. 4 people; a further 24 minutes.

● Exercises

1. The right-hand column should read

   1, 2, 3, 2, 3, 4, 4, 4, 4, 4, 5, 5, 5, 6, 7, 6, 7, 6, 7, 8.

2. The right-hand column should read

   1, 2, 3, 2, 3, 4, 3, 4, 4, 3, 4, 4, 4, 3, 3, 2, 3, 4, 3, 3.

8.4. Who is supposed to be able to read this?

TEACHING NOTES

This problem involves various peripheral skills in addition to arithmetic, which provides an opportunity to use the calculator memory, if desired. These other skills include counting and the use of tally marks, table-filling, simple formulae, use of a nomogram, and above all the constant need to ask "Is this a sensible answer"?

For further reading on testing for reading age see "Readability in the classroom" by C. Harrison.

At the start of the session *give them time* to read the two newspaper articles.

Basically the *Sun* is easier to read because it has shorter sentences and fewer 'long' words.

Get the students to write down their guesses as they will need these to help them assess the correctness of their calculated answers.

You may need to comment on the word 'syllable'. The students will find that they do not necessarily all agree on the length of words, as this depends on their own pronunciation.

For example us–u–al–ly   or us–u–ly.
1 2 3 4   1 2 3

This does not matter since these discrepancies have little effect on the overall formula.

It is worth showing the students some tally method e.g. ~~1111~~, ~~1111~~, 111, since it is much the easiest way to count large numbers of anything, and, for example, is useful when doing a stock control, or checking deliveries.

The snag attached to the Forecast formula is that it has a minimum value of 10, and so it is not usable on children's literature. The reason for this is that no account of sentence length is taken. It is, however, useful for this very reason, for assessing forms and documents which are not written in complete sentences. It is also very easy to use.

● Problem solutions

The figures used below are obtained from Table S8.1.

Table S8.1

Sentence number	*Sun* Words	Syllables 1	2	3	4	5	Total	*Observer* Words	Syllables 1	2	3	4	5	Total	*Child's story* Words	Syllables 1	2	3	4	5	Total
1	7	5	2	–	–	–	9	32	18	7	3	3	1	58	14	11	3	–	–	–	17
2	5	5	–	–	–	–	5	68*	42	17	6	2	1	107	22	19	3	–	–	–	25
								9	7	1	1	–	–								
3	15	12	1	2	–	–	20	41**	24						8	6	2	–	–	–	10
4	26	22	3	–	–	–	33								10	10	2	–	–	–	10
5	1	1	–	–	–	–	1								5	5	–	–	–	–	5
6	23	16	6	1	–	–	31								13	13	–	–	–	–	13
7	23*	17	5	1	–	–	30								10	9	–	1	–	–	12
8	15	10	3	1	1	–									11	11	–	–	–	–	11
9	6	5	–	1	–	–									7*	7	–	–	–	–	7
10	8	8	–	–	–	–									12	12	–	–	–	–	
11	11	6	4	1	–	–									10	9	1	–	–	–	
12	1	1	–	–	–	–									7	6	1	–	–	–	
13	9**	8	1	–	–	–									11	11	–	–	–	–	
14															9**	8	1	–	–	–	
Totals	150	116 (N)					119 (S)	150	91 (N)					165 (S)	150	137 (N)					110 (S)

* 100 words has been reached at this point.
** Not all of the words in this sentence have been used – only the number required to bring the total number of words to 150.

1. Reading age for *Sun* is $R = 13.4$ years.
2. Reading age for *Three pigs* is $R = 11.3$ years.
3. Maximum $N = 150$ and minimum $R = 25 - 150/10 = 10$. It would seem that, according to the forecast formula, there are no books suitable for children under 10 years old!
4. *Sun* $W = 141/12 = 11.75$.

   Observer $W = 109/2 = 54.50$.

   Three pigs $W = 141/13 = 10.85$.

NOTES ON METHODS IN 'RELATED PROBLEMS'

A. The FOG method gives a good spread of scores, but does not really cope with easy material. 'This is Sam. Sam is a dog. Sam likes meat.', etc. could produce a reading age of 1.2 years! It has the advantage that counting only polysyllabic words is fairly quick.

B. This method gives a very low age for the *Observer* and, in fact, has a limited range of use. The lowest age predictable is 7.4 years and because of the low weighting given to sentence length, the formula exhibits a ceiling effect. It is useful for primary school materials.

C. Flesch's method is the best known formula. It was originally designed for assessing adult materials, hence the need to convert the score to a reading age. It is also less discriminating at the lower age range.

This may be the student's first encounter with a nomogram, so a little explanation may be needed.

It is in fact not possible, except by projecting Flesch's scale beyond a score of 100, accurately to assess the child's passage.

● Related problems

A. *Sun:* $R = 0.4\,(11.75 + 6) = 7.1$ years.

Three pigs: $R = 0.4\,(10.85 + 0.67) = 4.6$ years.

B. *Sun:* $R = 2.7971 + (0.0778 \times 11.75) + (0.0455 \times 119)$
$= 9.1$ years

Observer: $R = 2.7971 + (0.0778 \times 54.50) + (0.0455 \times 165)$
$= 14.5$ years

Three pigs: $R = 2.7971 + (0.0778 \times 10.85) + (0.0455 \times 110)$
$= 8.7$ years.

C. For Flesch's formula we obtain the following

Sun: score = 94.2; $R = 5.6$ years.

Observer: score = 11.9; $R = 19.2$ years.

Three pigs: score = 104.07; $R = 4.7$ years

For Flesch's nomogram we obtain

Sun: score = 95.

Observer: score = 11.

Three pigs: score over 100.

8.5. Plugging gaps

A pre-metrication problem was to show that the best set of weights for any whole number of ounces consisted of

1, 2, 4, 8, . . . ounce weights.

Indeed this was the way that sets of weights were made up. Nowadays we can hardly use this very apt illustration of counting and representation in the binary scale. However, problems 1–3 are about a similar situation.

1.

Gap width	Blade thickness (X 0.05 mm)		
(X 0.05 mm)	1	2	4
1	√		
2		√	
3	√	√	
4			√
5	√		√
6		√	√
7	√	√	√

2. 0.05, 0.10, 0.20, 0.40 mm.

Gap width	Blade thickness (X 0.05 mm)			
(X 0.05 mm)	1	2	4	8
1	√			
2		√		
3	√	√		
4			√	
5	√		√	
6		√	√	
7	√	√	√	
8				√
9	√			√
10		√		√
11	√	√		√
12			√	√
13	√		√	√
14		√	√	√
15	√	√	√	√

3. (a) The thicknesses are obtained by repeated doubling

$$1, 2, 4, 8, \ldots \text{ units.}$$

(b) Students should spot that in the first column of the table in Problem 2 there are ticks in alternate rows; in the second column two ticks are followed by two gaps; in the third column four ticks are followed by four gaps; and so on.

(c) With n blades we can measure $(2^n - 1)$ widths.

● Related problems

A. No.

B. (a) 4; (b) 5; (c) 7; (d) 7; (e) 8.

C.

Weight measured (kg)	Weights used in measuring		
	1-kg	3-kg	9-kg
1	+	0	0
2	–	+	0
3	0	+	0
4	+	+	0
5	–	–	+
6	0	–	+
7	+	–	+
8	–	0	+
9	0	0	+
10	+	0	+
11	–	+	+
12	0	+	+
13	+	+	+

The patterns in this table should help the ambitious student who wants to look at the next case.

D. There is something special about 1, 3, and 9, of course. But 13 different weights could be made from other sets of three. However, they would not cover all the whole numbers varying from 1 to 13. A complete discussion on why the set $1 = 3^0, 3^1, 3^2, \ldots$ should be 'best' for this sort of weighing is out of place here, but it is illuminating to note that whereas in problems 1–3 there were only two choices (include or exclude) for each blade, in questions C and D there are *three* choices (add, subtract, or omit) from each weight.

The blacksmith should cut the bar as shown in Fig. S8.1 into pieces weighing 1, 3, 9, and 27 pounds.

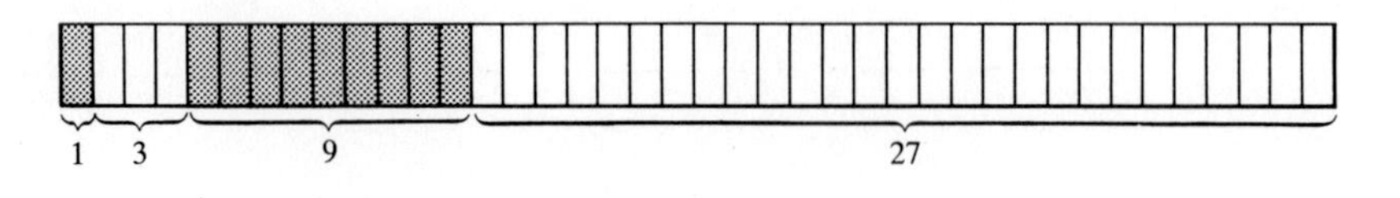

Fig. S8.1

8.6. Map colouring

The solutions to the problem are found in Fig. S8.3 (a)–(d). The key is B, blue; R, red; Y, yellow; G, green.

The solution for Africa uses four colours. One possible colouring is given in Fig. S8.2.

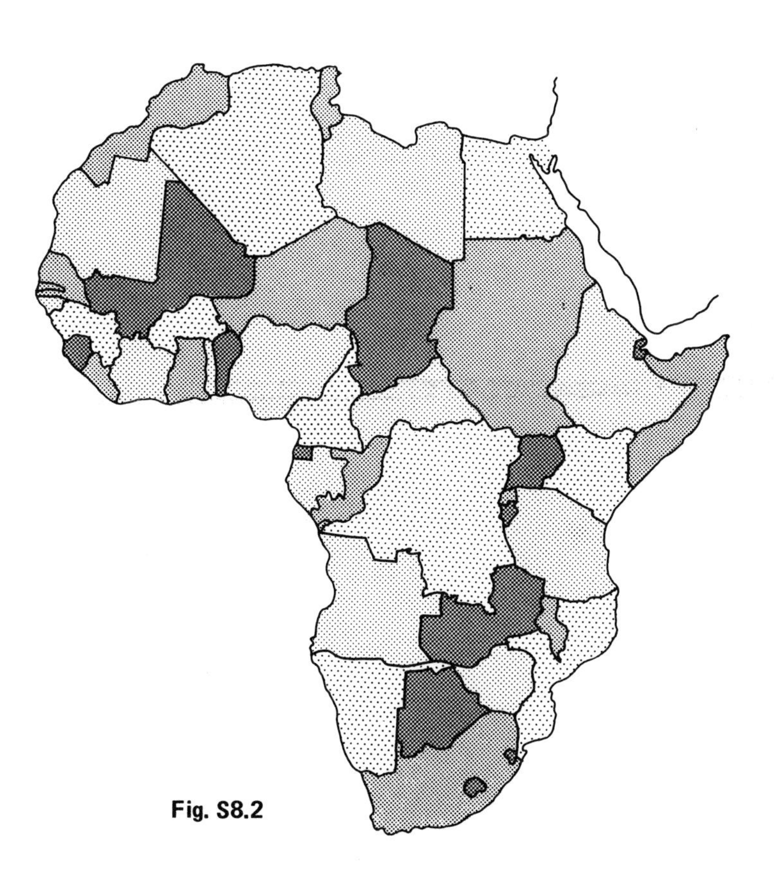

Fig. S8.2

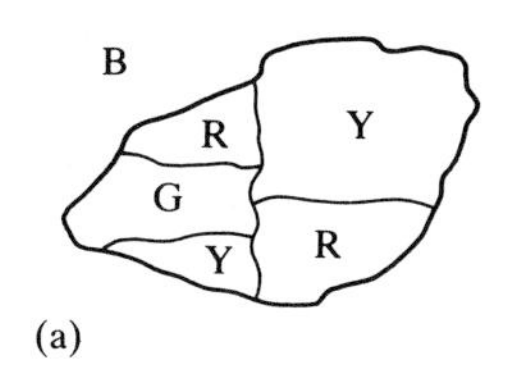

(a)

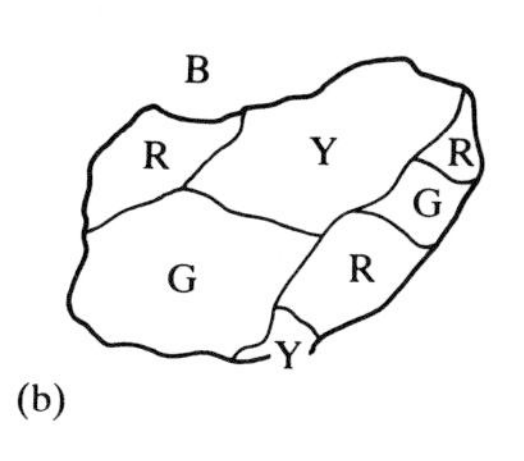

(b)

(e)

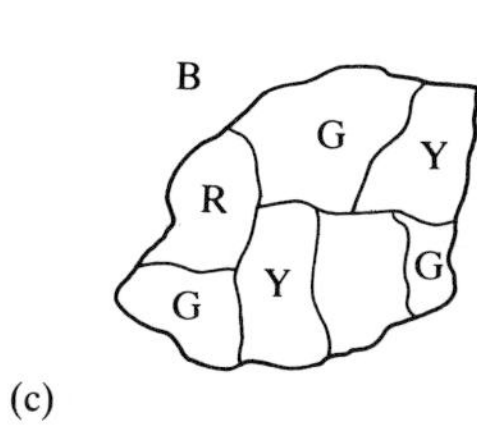

(c)

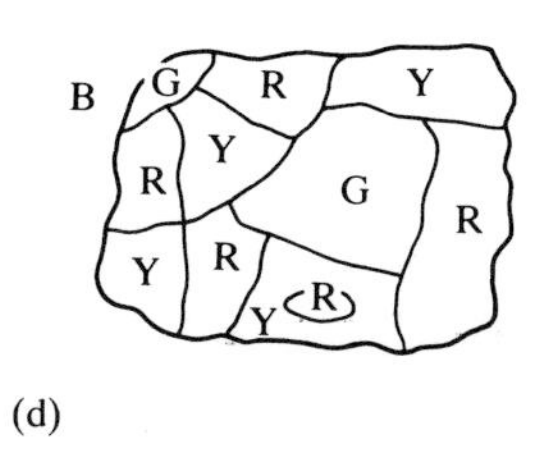

(d)

Fig. S8.3